The World of
QUANTUM CULTURE

The World of
QUANTUM CULTURE

Edited by Manuel J. Caro and
John W. Murphy

PRAEGER   Westport, Connecticut
London

Library of Congress Cataloging-in-Publication Data

The world of quantum culture / edited by Manuel J. Caro and John W. Murphy.
    p. cm.
  Includes bibliographical references and index.
  ISBN 0–275–97068–X (alk. paper)
  1. Culture—Philosophy. 2. Aesthetics. 3. Quantum theory. I. Caro, Manuel J., 1970– II. Murphy, John W.
  HM621.W65 2002
  306'.01—dc21   2001058062

British Library Cataloguing in Publication Data is available.

Copyright © 2002 by Manuel J. Caro and John W. Murphy

All rights reserved. No portion of this book may be reproduced, by any process or technique, without the express written consent of the publisher.

Library of Congress Catalog Card Number: 2001058062
ISBN: 0–275–97068–X

First published in 2002

Praeger Publishers, 88 Post Road West, Westport, CT 06881·
An imprint of Greenwood Publishing Group, Inc.
www.praeger.com

Printed in the United States of America

∞™

The paper used in this book complies with the Permanent Paper Standard issued by the National Information Standards Organization (Z39.48–1984).

10 9 8 7 6 5 4 3 2 1

I would like to dedicate this volume and all the work behind it to my family (nuclear and extended) and to the Quantum Aesthetics Group.

—Manuel J. Caro

Contents

Preface	ix
Chapter 1 Overcoming the Limit Syndrome, *Gregorio Morales*	1
Chapter 2 Quantum Language, *Mihaela Dvorac*	35
Chapter 3 Quantum Literature, *Francisco Javier Peñas-Bermejo*	47
Chapter 4 Aesthetics as Creation and Transformation of World Awareness, *Algis Mickunas*	71
Chapter 5 Quantum Aesthetics and Art History, *Jennifer Wilson*	89
Chapter 6 Quantum Aesthetics: Art and Physics, *María Caro and Andrés Monteagudo*	111
Chapter 7 Individuation and Social Group, *Juan Antonio Díaz de Rada*	127
Chapter 8 The Interpretative Foundations of Culture: Quantum Aesthetics and Anthropology, *Graciela Elizabeth Bergallo*	145

Chapter 9
Quantum Aesthetics and the Polity, *John W. Murphy and Manuel J. Caro* — 165

Chapter 10
Conclusion: The Renewal of Cultural Studies, *John W. Murphy and Manuel J. Caro* — 181

Suggested Readings — 189

Index — 193

About the Contributors — 201

Preface

At the beginning of 1999, María Caro asked for our help in translating into English a manifesto that a group of her fellow artists and writers had put together. Since that moment, we became interested in the project of quantum aesthetics. With roots in both phenomenology and postmodernism, John W. Murphy had been working for years in the development of a theoretical framework for sociology and cultural studies that broke away from dualism; a dualism that had harmed Western thought, as well as other people and cultures. Manuel J. Caro joined this search a few years ago, and together we started the project whose final product you are about to read.

Quantum aesthetics represents for us a new way of approaching the world that is very compatible with our philosophical convictions, even though it has different bases. For some time, our antidualist work had been derived from the humanities and philosophy, although now we face a humanist and cultural project that is antidualist and based on physics—the epitome of science. Since quantum aesthetics is predicated on ideas that originate from the material realm of reality, as opposed to the immaterial domain (a critique that has been lodged against the humanities), any theory of culture that emanates from this theoretical framework should have the legitimacy that positive science has enjoyed. Furthermore, this project should surpass positive science for it breaks away from the untenable dualism and realism that has plagued science since the Renaissance. Indeed, the separation between the physical and immaterial world is shown to be false by quantum aesthetics: intuition and matter are integrated in an undifferentiated whole.

We devote this volume to the development of an antidualist model of art, society, and culture. This might not be the first book that deals with quantum aesthetics or with the consequences that the discoveries of subatomic physics have for the humanities and social sciences. However, this is the first book that deals with all these themes at the same time and with the explicit intention of creating a comprehensive theoretical framework for cultural studies. This *is* also the first book that includes a variety of different topics and authors from inside and outside of the Quantum Aesthetics Group. We are sure that this volume will make an important contribution to the dialogue among those who are interested in such differing topics as identity, the polity, art, society, history, and language.

The first chapter, written by Gregorio Morales, is an introduction to the central themes and concepts of quantum physics and to the consequences that these ideas have for art, literature, and the humanities. Chapters 2 and 3, written by Mihaela Dvorac and Francisco Javier Peñas-Bermejo, respectively, deal with the consequences of subatomic physics for language and literature.

The focus of Chapters 4, 5, and 6 is the world of art. Algis Mickunas discusses the crucial role of aesthetics in the creation of reality persons know and experience. He also illustrates the type of world that quantum aesthetics helps to produce. In Chapter 5, Jennifer Wilson delves into the basic principles of quantum art and locates them within the history of art, while in Chapter 6 María Caro and Andrés Monteagudo write about both the artists of the Quantum Aesthetics Group and their work.

Chapters 7, 8, and 9 explore the implications of quantum principles for the social sciences. In Chapter 7, Juan Antonio Díaz de Rada reveals the importance that Carl G. Jung's concept of individuation has for quantum aesthetics, as well as how this concept leads to the understanding that individuals are not atoms who are, at best, only indirectly a part of a community. Graciela Elizabeth Bergallo writes in Chapter 8 about the quantum concept of culture and what this outlook means for anthropology. In Chapter 9, John W. Murphy and Manuel J. Caro sketch some of the consequences that taking the antidualist ideas of quantum aesthetics would have for the construction of society. Finally, in Chapter 10, John W. Murphy and Manuel J. Caro review the most important ideas raised by the authors in the various chapters, as well as highlight the possibilities that the theoretical framework discussed in this volume presents for constituting a comprehensive theory of cultural studies.

Before you begin to read the book, we would like to clarify an important point. This book has been compiled from original manuscripts written in both Spanish and English. Thus, this volume includes some translations, not only of the chapters themselves, but also of quotes and references to third authors. So, if in a footnote you see a reference in Spanish, even though the text of the quote is in English, you should always keep in mind that the original was translated into English by Luigi Esposito, Maritza Flores, Leeann Hunter, Manuel J. Caro, or John W. Murphy. We have decided not to follow the usual practice, well established in academics, of leaving the quotes in their original language, because we did not want to lose those readers who might not be able to read these languages. We want to take this opportunity to thank both the authors and other translators who helped us make this collection a reality.

CHAPTER 1

Overcoming the Limit Syndrome[1]

Gregorio Morales

THE RESIDENTS OF THE SMALL HOUSE AND THE BLACK SHEEP

Once upon a time, there was a group of petite bourgeoisie who, afraid of the world that surrounded them, built a comfortable little house, hid themselves inside the small structure, locked all entrances, windows included, and declared that everything that existed outside the house was metaphysical. Wanting to uncover the laws of reality that they proclaimed resided inside the house, these individuals devoted themselves to studying every feature of the residence, including the walls, floors, and woodwork. Everything functioned like a perfect mechanism whose patterns became satisfactorily identified and established by the residents. If one day there happened to be a storm and a thunder was heard, or spurts of luminosity were seen and residents asked what this was and why this was taking place, they were told that these questions made no sense and were reproached for the sake of discouraging such questions in the future. "Whatever resides out there cannot be known," claimed the leader of the residents. "Asking such questions relegates us to a state of infancy. We should not question why, but rather how." The residents of the house continued their daily affairs in an orderly, systematic way, establishing laws that were precise, harmonious, and clear, all of which were adequately tested and verified. "Because all natural laws are invariable," argued the group's leader, "every time we discover one of the laws, we can foresee its future behavior." The leader added: "These laws are valid for all men and all historical epochs." The residents of the house applied these same sorts of principles to language, establishing a dichotomy between real and false judgements. "The limits of the house coincide with those of language," claimed the leader. Accordingly, asking questions that emerged from one's mind was also considered something metaphysical, a senseless fantasy. In this manner, the moment arrived when these individuals talked exclusively about the house and what they knew about the house, and, by extension, about their bodies and what they knew about their bodies. They could say "I am defecating," because such a statement was empirically testable. But they could not state "I think," because this idea eluded the principles of verification that they had established.

One day, one of the residents was walking along the lustrous parquet in the living room when, all of a sudden, one of the floor's wooden slabs "jumped" toward him. Later another wooden slab did the same, then another and another! Astonished because this had never been seen before, and the laws of the parquetry said that the floor should remain firm for many years, the resident dissented by asking: "Why has this occurred?" The leader heard the resident's question and reproved him: "That's a metaphysical question!" Following the leader's reprimand, his scientists spread out and diligently began to study the case. During the course of many days of close observation, they concluded that the wooden floor had slightly increased in volume and this change had caused the wooden slabs to be displaced. But the poor black sheep (i.e., the dissident resident) continued to ask questions, although now furtively: "Why has the wood increased in volume?"

Clandestinely, the black sheep undertook an investigation and discovered that the wooden floor was humid. For many years, he continued to study a hole that had formed at the center of the living room and found that, in the foundation of the house, there occurred dry and humid periods. It was during the latter that the wooden slabs distend and thus "jump." "Therefore," this pariah concluded, "humidity must have influenced this distension." The black sheep informed his fellow residents, who angrily responded: "The reason you propose lies outside of the house. But every explanation must be found inside!" Nonetheless, the heterodox became fascinated with the unfathomable abyss that lay before him and resumed his secret investigation, as he rejected the formal rules of the house. He did so to prove that, during the humid periods, the house tended to be hit noisily from above in hundreds of different places, while during the dry periods the house remained silent. The black sheep concluded that the humidity comes from the sky, hits the house, and later settles on the ground, where it infiltrates the parquet. Later, the humidity gradually disappears during the dry periods, when the floor's wooden slabs return to their original size. Unable to keep his excitement a secret, the black sheep once again informed his fellow residents of his findings. The house leader, no longer able to bear with the black sheep's scandalous conclusions, asserted: "That is no longer simply metaphysical, that is Nazism!" Tearing at his own clothes from anger, the leader yelled: "From the sky! We cannot reproduce something that comes from the sky in a laboratory! Besides, something cannot simply appear and vanish just like that." Fed up with the black sheep's outlandishness, the leader condemns him to death. The heterodox, at the brink of his own demise, becomes increasingly lucid and cheerfully exclaims: "I got it! I got it! Water, humidity, and vapor are the same thing." At that moment, his fellow residents incinerate him. "What?" the residents cry out in unison while the hapless black sheep agonizes in the flames. "Various things are the same thing? 'A' and 'not-A' can never exist at the same time."

NAIVE POSITIVISM AND ASPECT'S EXPERIMENT

Narrations such as the previous one inevitably come to mind when I read any of the classic texts of positivism. These texts are tales informed by naive positivists (the "ingenuous") who, consistent with Friedrich Schiller's wise distinction, are confused and surprised by the object.[2] They are possessed by the limit syndrome; that is, they crash against the four walls that surround them. "Limits can never be transcended," claims the naive positivist, who has imposed his overwhelming dominance throughout most of the nineteenth and twentieth centuries. Based on judgements of these naive thinkers, the belief in universal principles that exist independent of subjective value become reified, and, as a result, all other ways of understanding the world that deviate from these rules automatically become identified as false or irrational. These sorts of considerations, almost always implicit in any theory, are typically expressed by the term "dualism": the laws of science and the truths they represent are universal and transcend historical periods and cultures. Through this theoretical maneuver, Western culture is established as "the Culture," while all other cultural expressions are evaluated according to this reified knowledge base and devalued. Based on this idea, Western societies reserved for themselves the right to colonize and civilize any other culture.

But quantum physics has wounded the naive positivist at the center of his very being, and, with him, all of positivism. According to a famous experiment undertaken by Alain Aspect (1982), a very large percentage of the polarization angles of photons emitted by a laser beam are identical.[3] This means that the particles necessarily communicate their position, so that each photon's orientation can parallel that of the one that serves as its pair. But if this is the case, this communication is instantaneous. John S. Bell's theorem supports this conclusion by negating the idea that the world is local, thereby allowing simultaneous actions across space. Our impenitent heterodox would conclude, in this regard, that there must necessarily exist a type of communication superior to the speed of light. Reputed scientists have also affirmed this fact. For example, subsequent to Aspect's experiment, Brian Joseph, Nobel Prize winner in 1974, recognized that "this fact implies the possibility that one part of the universe may have an understanding of another part, or that two parts can come into contact under certain conditions independent of distance."[4] However, the confused positivist, who continues to fail to conclude the obvious, claims: "The experiment may indeed be true. But there can never be a speed superior to the speed of light! Never!"[5] To paraphrase Antonio Machado, rejecting anything that escapes their sensible pieces of evidence, positivists "despise what they ignore."[6] Moritz Schlick, the founder of the Vienna Circle, affirms without embarrassment that empiricists do not contradict metaphysicists; instead, "they say I don't understand you." In other words, positivism is nothing but the confirmation of impotence.

But the heterodox researcher has already opened a great abyss in the peaceful and comfortable passage of time, and this divide will gradually undermine the absolutes adhered to by those who remain inside the house. As each of these absolutes is exposed, the leader, who was adequately trained, continues to claim: "That is a lie!" But before his frightful eyes, his dogmatic assumptions about the independence between the observer and the observed (i.e., dualism), about scientific objectivity and its treatment of the world of senses as something given and inherently real (i.e., foundationalism), about principles of causality, and about Aristotle's proposition that "A" and "not-A" can never exist at the same time are all destroyed. In this manner, science ceases being eternal, immutable, and simple, and becomes temporal, fluid, and complex.

What is perhaps most interesting about this—and this is what is crucial about our story—is that the strict, rigorous principles of positivism are precisely what has led to its epistemological collapse. Our poor black sheep was, in reality, the best physicist of all the residents. His questions were the most relevant and insightful, and they led to a discovery that may have been easily verified. The bad news is that each of his findings was equivalent to placing a bomb in the foundation of the little house.

BOMBS AGAINST THE LITTLE HOUSE

Let us make an inventory of these dangerous "bombs." The majority of these derive from subatomic physics; the rest from related sciences, such as astronomy and biology.

Principle of Complementarity

Depending on how we measure it, a corpuscle can behave as a wave or a particle. This idea can be expressed by stating that a corpuscle is both a wave and a particle, although both cannot be observed at any single measurement.

Principle of Uncertainty

It is impossible to determine the position and momentum of a particle through a single measurement. Evidently, both occur at the same time. But to measure one, we have to retain an element of uncertainty with regard to the other, thus breaking from the sort of certainties proposed by Isaac Newton.

Anthropic Principle

The observer modifies the experiment with his/her observations. This is known as the "Hawthorne Effect." This phrase is derived from a research project, undertaken between 1927 and 1929, designed to compare levels of productivity between two groups of employees at the Western Electric Company in Chicago. One group was given work breaks, while the other was not. What the researchers found was that, independent of their group affiliation, the productivity of workers increased by 30 percent when they were being observed.[7] Therefore, the experiment itself affected the study! And this sort of effect is produced at all levels of human inquiry.

Moreover, all equations and theorems of the physical world are suspiciously similar to humans. That is, in the process of studying any subject matter, we inevitably encounter a mirror that reflects our own image. Erwin Schrödinger's theory of the "parallel universe" is consistent with this idea: There are hundreds of universes that surround us, but it is the observer who gives any particular one its observable form. Bereft of an observer, our universe would simply be one of many parallel universes that flow in the cosmos. Because reality inevitably requires observation and recognition, the human element necessarily mediates its meaning. Once the tie between human praxis and knowledge is taken seriously, the idea of scientific objectivity becomes questionable, and, as already discussed, dualism and foundationalism are undermined. Reality is no longer simply something that exists independent of humans; instead, reality is the product of the relationship between the millions of possibilities and our own subjectivity.

Nonseparability

People tend to view the universe as a composite of discrete objects. However, according to recent equations and experiments, this is merely an illusion. The universe seems to be a unified whole that is interrelated and devoid of limits (not local), where, to borrow from Francis Thompson's poetic expression, "one cannot uproot a flower without disturbing the stars."[8] In other words, we are connected to everything and everyone. The theory of "superstrings" offers a plausible explanation of this idea by claiming that the universe comprises different and interconnected vibratory components, much like the neurons of a brain. Pristine and autonomous laws, in short, do not exist. Rather, all laws are intertwined with one another and thus operate through their multiple interactions.

Acausality

In the everyday world, we typically assume that any phenomon is always and necessarily caused by some other event. However, in the subatomic world, this is not necessarily true, for acausal facts exist. For example, some scientists have found that atoms appear and disappear through a process that is spontaneous and not causal. This undermines the principle of causality that says every effect is preceded by a cause. The world and all matter are constantly created from nothing.

Complexity

Although everything in the cosmos obeys the law of entropy, there is nonetheless a tendency among organisms—from the particle to both cells and interstellar gasses—to organize themselves into increasingly complex structures. Accordingly, any research that studies the order of systems must take into account what has been termed "complex thinking."[9] Any investigation that fails to do so results in an oversimplification of the reality that surrounds us and thus is considered to be puerile. For example, by overlooking all the exterior causes that influenced the internal laws of the house, the scientific system adopted by the residents was guilty of this sort of silliness.

Ubiquity

Whoever thinks that particles can be found in any one concrete locality is wrong. Instead, particles behave in such a way that they can be found in many places at once, as if they had a ubiquitous quality that allowed them to be "conscious" of phenomena that may take place far away, anticipate obstacles, and advance responses.

Morphogenic Fields

Very likely, the existing connection between all points of the cosmos brings about specific standards, indicating that, for example, there exists a universal spirit or a spirit of a specific epoch. Moreover, we can take this a step further and see in specific connections the instincts of a species, including its behavioral and learning guidelines. That is, these morphogenic fields represent modalities of structures, rules, behaviors, ideas and tendencies, each one informing particular aspects of reality. The biologist Rupert Sheldrake came up with this revolutionary concept and tested this idea in his book *Seven Experiments that Could Change the World*.[10] In his opinion, one of these morphogenetic fields consists of the accumulated experiences of humankind. As a result, any changes or novel actions initially collide with strong sources of resistance; however, as these changes become adopted by an increasing number of people, learning them becomes easier. For this reason, morphogenetic fields, at least as they relate to humanity (and other superior animals), are ever-changing and in a process of continuous transformation. Each action exists within the context of a morphogenic field, incorporating itself into the whole while simultaneously changing this composite.

The Existence of "A" and "Not-A" at the Same Time

The arguments raised in the concept of "fuzzy thinking"[11] leave no doubt with respect to this point. Starting in the 1970s, Lofti Zadeh developed a new basis for viewing reality.[12] His notion of fuzzy thinking casts doubt on conventional mathematics and especially the typical method of calculating probabilities. Based on fuzzy logic, we can create intelligent machines that actually learn as they are used. For example, take a bite out of an apple, and the resulting object is at the same time an apple and not an apple. What is important to know is the percentage represented by each phase of the apple. Initially, the probability of the object being an apple is high. But as we eat the fruit, the object in question is reduced as the "nonapple" factor increases. This represents such an obvious truth that fuzzy mathematics is very easy to comprehend.

Beauty as the Corroboration of the Relevance of a Theory

The beauty, harmony, order, and simplicity of a hypothesis are considered to be proof of its certainty. Indeed, many scientists are not willing to accept a principle that is not beautiful. This beauty resides, for instance, in the recent theory of "superstrings," which, on the one hand, is making understandable some of the paradoxes raised in quantum physics, and, on the other hand, is linking macrocosmic and microcosmic theories. The physicist Brian Greene refers to this type of beauty as an "elegant universe."[13]

The Universe as a Hologram

For the majority of scientists, the universe constitutes information. Moreover, this information—from the brain to the cells that comprise our body—is holographic; that is, each part contains the whole.

Manifest and Implicate Order

The universe has two types of order. The first, a *manifest* order, appears before our eyes; the other, an *implied* order, is hidden from our sight and only revealed at specific moments. David Bohm illustrates this distinction in his famous example of the ink stain contained within a white fluid that is moved by a cylinder.[14] At one point, the ink stain is visually evident. However, if we move the cylinder the stain gradually disappears before our eyes. Has the stain disappeared? To be sure, the ink can be found "folded" inside the fluid. If we move the cylinder in the opposite direction, however, the stain gradually reappears until it reaches its original form.

No Distinction between Matter and Energy, Mind and Body

The idea that matter is energy is seldom questioned by any contemporary physicist. Furthermore, the idea that what has been referred to as "mind" or "spirit" is another manifestation of matter is progressively being accepted. Arthur Eddington, the director of the expedition that photographed a solar eclipse that confirmed Albert Einstein's theory of relativity, affirms: "[T]o put it bluntly, my conclusion is that the world is composed of mindful matter."[15]

This is not to say that matter is strictly mental, but rather that both mind and matter are manifestations of a common basis that, to this day, we do not fully understand. As suggested by Wolfgang Pauli, Nobel Prize winner in physics in 1945, "I prefer to say that both the psyche and matter are governed by a common order."[16]

THE THIRD WAY

With these discoveries, the small house has been completely obliterated, as if a devastating storm had passed over the house, showing that what existed outside was not at all metaphysical, but rather constituted a reality that is a lot more profound than is allowed by the restrictions imposed by the residents. Contrary to what the residents believed, breaking from these limits does not require that we invoke some sort of fantasy. Rather, reality is incalculably larger and deeper now. We no longer run the risk of falling into the trap of metaphysics by entertaining the possibility that, under certain circumstances, traveling through time is a feasible undertaking, that *all* of time is found in any given parcel of time—past, present, and future. Also, we do not fall into a metaphysical trap if we suggest that the mind can influence matter; that intelligence is no longer exclusive to living beings; that it is possible the most remote parts of the universe are able to communicate instantaneously with one another and we are able to communicate with them; that our minds are not sundered from one another but rather embedded within fields that are common to us all; that beauty is a property of the entire universe; and that humans are the ones that, through their observations, create the reality that we know.

One of the first disciplines that took on these new ideas was aesthetics, probably because this discipline deals with a fecund terrain where the most inextricable jungles can grow. However, paradoxically, the devastation brought about by positivist loggers has become today more apparent than ever before, converting literature, painting, sculpure, music, and film into the most desolate and trivial of pastures.

The twentieth century began with vanguard movements such as surrealism, which sought to vindicate imagination and fantasy that were marginalized by nineteenth-century realism. But the subsequent fascination with Marxism (which included many surrealists) and the fallacious aesthetic slogans of its theorists, together with the coarse invention of "socialist realism,"[17] soon constrained this wave of freedom. Socialist realism began to grow not only among the countries of the former Soviet Union and its satellite nations, where people who failed to adhere to its rules were tortured and assassinated in the basements of Lubianka,[18] but also in the West, where intellectual circles promulgated Soviet propaganda in a strangely effective way, equating progressivism with communism. The most distinguished European writers, artists, and intellectuals belonged, up until the late 1970s, to the Communist Party. Later, as a weak reaction against indolence and dull slogans, inane language games were imposed in the form of "experimentalism."[19] But such vacuity could only eventually collapse, and, once again, realisms made a successful return. In Europe, during the 1980s, socialist realism resurfaced. The name "dirty realism" was reserved for the most impoverished and imitative realism that was developed in North America.[20] And "magic realism" became the term for a folkloric type of realism that became widespread in Latin America.[21] A pure Newtonian world was contained within the four walls of the small house! Meanwhile, the subatomic world and the consequences of this domain were ignored by the great majority of all intellectuals, writers, artists, and politicians. The same sort of ignorance persists even among the most progressive of contemporary scientists. I recently heard one of these scientists talk about the great revolutions that quantum physics is initiating, when at one point someone asked: "What type of effect does all this have on our lives?" The investigator responded in a matter-of-fact tone: "None at all! The subatomic world is so small that it has no effect on our everyday lives." But these subatomic discoveries have changed every facet of the macroscopic world, starting with technology and moving on to art and literature, and on to philosophy, sociology, law, and politics! In this respect, the Spanish physicist Francisco J. Yunduráin affirms: "The transistor is the off-spring of quantum mechanics. The inner workings of television screens and computers are based on a quantum effect; moreover, the nuclear disintegration associated with the atomic bomb is also a quantum effect."[22] So, the subatomic world has no effect on our lives! Right!

Quantum physics has inaugurated a new paradigm and everything is shaped by this framework. For this reason, subatomic principles are promoting a new culture that, as is usually the case, is beginning to emerge within the realm of art. Leaving critics in awe, Gustave Flaubert's century-and-half-old prophecy has been realized: "Each day art will become more scientific while at the same time science will become more and more artistic. These two disciplines will rediscover themselves at the end, after having separated at the beginning."[23]

As a result of the debates between idealists and positivists, the arts and sciences were separated during Flaubert's time. The reciprocal contempt between the humanists and the scientists culminated in the terrible and bloody division that is well known today. This division continues to be perpetuated not only because of differences in training, but also because humanists do not trust science and scientists tend to feel the same way about the humanities. This mistrust is still significant nowadays: Humanists consider modern science to be apocalyptic and inhumane, and scientists consider the humanities to be nothing but a series of value claims devoid of rigor. Given this feud, it is not surprising that when we speak about *quantum aesthetics*, people tend to imagine an art that is cold, indifferent, and mechanical, when in reality it is quite the opposite.

Now, however, we are at the beginning of a period where the humanities and sciences are converging thanks to the increasingly general recognition that both are valid ways of knowing the world. Science cannot help but to enter the domain of art and literature in order to develop concepts that are indispensable to scientific inquiry, just like art and literature must enter the realm of science to get in touch with the intuitions and fantasies that form a part of perennial wisdom, and, in this way, be able to understand humanity in a more ample and profound way. Art and science are also brought together by the fact that both share beauty as their yardstick and ultimate goal.

According to Leonard Shlain, who has studied in-depth the relation between physics and art, the latter has always been many decades more advanced in terms of subatomic discoveries, showing in an artistic way what later has been converted into principles and equations. In this respect, Shlain affirms:

And here the thesis of this book—that revolutionary art anticipates visionary physics—lyes revealed. When the vision of the revolutionary artist, rooted in the Dionysian right hemisphere, combines with precognition, art will prophesy the future conception of reality. The artist introduces a new way to think about the world, then the physicist formulates a new way to think about the world. Only later do the other members of the civilization incorporate this novel view into all aspects of the culture.[24]

Quantum aesthetics has been one of the first movements that deliberately and consciously has sought to promote the reunification of the sciences and humanities. Indeed, this tendency has become unavoidable. Even scientists such as Edward O. Wilson, who work within the most adamant positivism, advocate this union. In this regard, Wilson has elaborated this point with his concept of "consilience."[25] The "[m]ind's major intention," he affirms, "has always been and will always be the intention to link the sciences with the humanities."[26] He adds: "Neither the sciences nor the arts can ever be complete without combining their respective strengths. Science requires the intuition and metaphorical power of the arts, and the arts require the new blood of the sciences."[27]

Quantum aesthetics opens a third way in the arts that is, on the one hand, divorced from nineteenth-century realisms including socialist realism, and, on the other hand, distanced from *pompier* fantasy (or the combination of both, which constitutes magical realism). This mode of aesthetics deals with a reality whose apparent limits can be transcended without falling into anything phantasmagorical. The stories of the Argentine José Gabriel Ceballos provide an excellent example of this third way.[28]

The revolutionary task of quantum aesthetics is radical in that there is no need to exit the realm of the sciences to inaugurate a paradigm that undermines the old canons, while fostering scandals and freedom. That is to say, if it was science that constrained aesthetics within realism in the eighteenth century, then it is science that restores to art the grandeur, possibilities, and mysteries that were lost. This perspective is completely novel and its consequences have never been put into practice. In this regard, an immense task lies before the quantum thinker: that of reconstructing the reality in which we live from the very conceptions that have shattered the limits of the small house, while explaining, for perhaps the first time in our lives and in a very real way, that we can make our own destiny and we are not alone in this task. Additionally, quantum aesthetics illustrates that the actions of our ancestors and our descendants pulsate in us, and that we communicate with the entire universe. Also following the collapse of dualism, intuition can attain truth, coincidences are nothing but destiny embedded within our minds, and complexity operates in everyone's life. Furthermore, we are dead and alive at the same time—we are both mortal and immortal simultaneously—and there exists an awful and terrifying jungle of reality that is hidden from view. One of the main goals of our lives is to understand this reality. As I have stated, we have to reinterpret everything from the perspective of this new reality that has now been opened to us; we must write a new *human comedy* where the interrelations between parallel universes and their mysterious manifestations flow into our everyday reality.

A QUANTUM ART AND PSYCHOLOGY

In his letter to Louise Colet, Flaubert added to his diagnosis and intuitions about art that "human thought is not capable of foreseeing, in these moments, what resplendent psychological suns future art works will open."[29]

Was Flaubert wrong in speaking of "psychological suns," that is, of the relationships of both art and science with the psychological? Paul Davies affirms that "one of the principal victims of reductionistic sciences was the mind. . . . The new physics, in contrast, gives the mind a central position in nature." Referring to the surprising fact that the spin of a particle[30] always orients itself in the direction of reference indicated by the experimenter, Davies also writes that this issue introduces "a strange subjective element in the physical world. If the spin of a particle is destined to always follow the arbitrary lead of the experimenter, then, in some way, the free will of the experimenter inundates the micro world."[31] This, in his opinion, "seems to suggest that the mind dominates matter."[32]

Roger Penrose explicitly highlights this idea when he states that "physics may help us understand consciousness." And, to clarify, he adds: "I have the feeling that a new theory that unites the physics of the very large with that of the very small will have much to say about consciousness."[33]

Thus, it seems as if Flaubert was by no means mistaken! In this case, his prophecy has come true. The mind—according to psychologism—plays an essential role in the new paradigm. After all, how can we perceive the subatomic world in our everyday lives if not through a psychological process? The microworld influences our mind, that is, the way we view the world, including our perceptions, understanding, thinking process, and imagination. What is revolutionary about this mode of aesthetics is that, as in quantum physics, it goes beyond the physical realm of reality; the role of the mind and the observer is recognized in all quantum discoveries, and thus all findings are translated into the realm of psychology. This step is crucial if we are going to do away with the naivete of the residents of the small house.

Unfortunately, there are scholars who do not share or even understand this outlook. Nonetheless, this view is not only crucial, but absolutely normal. This is so because, if we agree that our instruments of observation influence whatever we are observing, how can we not take the mind—our most important instrument of observation—into serious consideration? Believing that whatever happens outside ourselves is objective and has no relation to the mind is to fall, once again, into the trap of naivete that we have criticized earlier. The mind participates more than any other *apparatus* in quantum freedom, but, at the same time, the mind conditions this liberty by making freedom conform to its own structure. How, then, can we neglect the mind? If the mind is important for physics, then it is vital for aesthetics. This change represents a pioneering activity, for any study of the mind has always been considered a mere fantasy. This is what Jorge Luis Borges had in mind when he ascribed religion and philosophy to the realm of fantastic literature.[34] However, subsequent to subatomic physics there is no way of escaping the idea that *whatever is mental is also real*. This view had already been adopted by Juan Eduardo Circlot, a poet and critic who was ahead of his time and considered to be one of the forerunners of quantum aesthetics. Circlot is also consistent with Giordano Bruno and Gérard de Nerval in the idea that thought and reality are intertwined.[35]

But is there a strand of psychology that directly responds to the issues raised by subatomic physics? If not, then such a discipline would have to be created, for the future requires this type of psychology. Fortunately, however, this psychology does exist and is none other than that developed by Carl G. Jung. Jungian psychology and subatomic physics have a deeply intertwined identity, one that was consciously sought by the Swiss psychologist. This relationship between subatomic physics and psychology was also emphasized by some of Jung's students, such as Aniela Jaffé[36] and M. L. von Franz. Franz affirms that "Jung ... discovered that, as a result of certain research projects, analytical psychology was required to create concepts that, later, became surprisingly analogous to those created by physicists when they studied micro-physical phenomena."[37] And later on, she claims: "The unexpected parallels in the ideas of psychology and physics suggest, as noted by Jung, a possible and definitive union between both realms of reality ... that is to say, there exists a *physico-psychological* union in all phenomena. Jung, moreover, was convinced that what he called the unconscious was intertwined, in some way, with the structure of inorganic matter."[38]

This union between psychology and physics motivated Jung and Pauli to reflect on and compare their ideas in an extended correspondence that lasted from 1932 to 1958. In these letters, one can observe in detail the progressive interpenetration of both theories.[39] Pauli writes to Jung: "[I]t is also very significant that many other important concepts such as equality (affinity), acausality, orderliness, correspondence, opposite pairs, and totality are applied simultaneously in psychology and physics."[40]

These sorts of affirmations provide merely a glimpse at the continuous reflections and proposals that Jung and Pauli made about the identity of physics and psychology. Today, after many years, this identity seems clearer to us than to their contemporaries.

THE MIND AS AN INSTRUMENT OF OBSERVATION

According to the directives proposed in the *uncertainty* and *anthropic principles*, it becomes impossible from a psychological point of view to enter our minds without modifying them. In this respect, whatever we say about our minds can never be an absolute or atemporal truth, but rather an approximation. We can never encompass, through narrative or art, a whole personality with its conscious and unconscious dimensions, but merely discover pieces that are always filtered by our instruments of observation, which are nothing more than an extension of our subjectivity. Humans are even more inscrutable than matter, more enigmatic than the universe, thus leaving behind the "one-dimensional man"[41] created during the twentieth century. Humans are neither reducible to what they consume (consumerism) nor to their behaviors (behaviorism), and much less can their intelligence be reduced to the parameters of a test. Such views are nothing but the sort of chicanery that the "wise" men of the small house employed to deceive the other inhabitants. Daniel Goleman has already begun to break from these perfidious illusions with his concept of "emotional intelligence."[42]

PARTICLES AND HUMANS

The British playwright Michael Frayn introduces this concept of the mulidimensional man in his work *Copenhagen*.[43] In this play, he raises the question of whether or not Werner Heisenberg attempted to build the atomic bomb for the Nazis. Frayn identifies the behavior of subatomic particles with his characters. In this way, in much the same manner that these particles have the ability of being in various places at the same time (ubiquity), he narrates different stories in which his characters play different roles. There is no absolute truth; instead, the truth unfolds always with uncertainty. In the author's words, Heisenberg's principle "states that an irreducible quantum fuzziness, caused by the wavelike nature of matter on small scales, makes it impossible to know simultaneously both a particle's position and its momentum. '*Humans' intentions have their own irreducible fuzziness.*' "[44] As the author states, applying his own principle to Heisenberg: "[T]he uncertainty about Heisenberg may be eternal."[45] The same thing goes on in our own lives, and art and literature must show this uncertainty. A biography is neither limited nor pristine, like the great works of Honoré de Balzac. Instead, a biography is malleable, interrelated with other biographies, contradictory, and interpreted by the observer. Evidently, this claim leads us directly to the notion of fuzzy thinking.[46]

THE OBSERVER AND THE PARALLEL UNIVERSE(S)

As we have seen, people's observations make reality what it is. That is, we *filter this universe out of* the infinite number of other parallel universes. Thus, the human being returns to occupy the center of the cosmos. This allows us to affirm that humans themselves "create" reality! This is precisely what takes place in Xaverio's "petrales."[47] On the one hand, if we stop in front of them, they seem homogeneous, like pure and simple surfaces with a permanent structure. On the other hand, if we move around them, everything we see in front of us becomes agitated, and our initial image of fixed surfaces turns into something that changes. This suggests that the moving spectator creates his/her own art. For this reason, Xaverio titled one of his works *Colores para Pasear* (*Walking Colors, or Colors for Strolling*).[48] These colors remain hidden and do not unfold until the viewer begins to move, after which fabulous rainbows emerge before the viewer's eyes. Devoid of a viewer, Xaverio's work would sadly be forced to keep its beautiful secrets.

Paul Auster's *Leviathan* offers the same sort of idea through literary expression.[49] Transformed as narrator and character in Peter Aaron, Auster reconstructs in *Leviathan* the life of his friend Benjamin Sachs, starting with the mysterious explosion that ended his life. Aaron is aware that each character constructs his own understanding of reality. Thus, far from imposing his own opinions as truths, Aaron always says "in the mind of," "in his version of history," or "to the point where I can judge him." In other words, as simply another character, Aaron offers his own version of the facts, but, at the same time, he respectfully presents how these same facts are viewed by others. Most incredible is that, in general, two different narrations are considered to be true at the same time. For example, at one juncture, the narrator is perplexed by two contradictory versions that Sachs has given him about his accident (he has fallen from a tall building) and comments: "I cannot believe both things. It has to be one or the other." But Sachs responds: "[B]oth are true."[50] This sort of "democracy" of viewpoints, or presentation of "parallel realities," occurs throughout the novel. "A" and "not-A" can exist at the same time!

While a particle can be in different places at once (ubiquity), the same is true of thinking. In much the same way that a part can account for the whole, a man can account for all of humanity and even everything that has been created. In this manner, what at first seems to be merely another detective story, later turns into something different once the narrator realizes that his own life is tied to those of the other characters. Following this insight, at the beginning of Chapter 2, he comes to affirm that "[i]f it were not for the break up of my marriage with Delia Bond, I would have never met Maria Turner, and if I would not have met Maria Turner, I would have never found out about the existence of Lillian Stern, and if would not have found out about the existence of Lillian Stern, I would not be sitting here writing this book."[51] In sum, Auster's characters live according to the principle of nonseparability. The destiny of any one character is tied to the destinies of all the others.

Although holographic theories had not yet been elaborated during Circlot's life time, and Aspect's experiment had not yet proved the inseparability of the universe, Circlot had already assumed in his work that the entire universe can be concentrated within the smallest of places and that the closest and most insignificant objects can lead to those that are most distant and unknowable. In *Los Restos Negros* (*The Black Remains*), Circlot shows how the past, present, and future simultaneously reside within each of us.[52] The universe is, therefore, an inseparable magma whose parts, as Charles Baudelare wisely perceives, are "confounded in a tenebrous and profound unity," where "smells, colors, and sounds respond to one another,"[53] establishing an infinite field of exchanges. Circlot affirms: "Correspondances? The art of the Irish monks during the time before and during the *Carolingian* Period is an art based on intricate bows, on labyrinths that are indefinite and absolute. This is the best proof that in 'this world'—this should be its ultimate justification—everything corresponds, intertwines, and communicates."[54]

For Auster, any person carries holographically within him/her the destiny of all of humankind. For this reason, he states: "Everything is related to everything else, every history overlaps with all others. However horrible it is to say this, I understand now that I am [that is, the narrator, Peter Aaron] the one who united us all. Just like Sachs himself, I am the point where everything begins."[55]

We are all like Peter Aaron, "the point where everything begins," which certifies our central role in the universe. Similar to the way in which contemporary astronomers consider any point in the cosmos to be the center, every person is also at the center of the cosmos. Accordingly, everything that surrounds us, from the closest to the most distant, is organized from within each one of us.

Something very similar to this occurs in the movie *Los Amantes del Círculo Polar* (*The Lovers of the Polar Circle*) by Julio Medem.[56] This movie narrates the love between Ana and Otto from the time they were eight years old until they were twenty-five. The words of Otto in the beginning of the movie resemble those of Peter Aaron. A school friend of Otto unintentionally kicks a ball outside the school yard and Otto runs after it; in that moment, in a nearby park, he encounters a young girl (Ana) who stares into his eyes. Otto's "off" voice reflects in the following manner: "[I]f she had stared at me less, I would have continued running to retrieve the ball; or, a moment earlier, if the kid who kicked the ball outside of the schoolyard would have kicked it more accurately, the ball would have stayed inside of the yard; and I would have been at the goal had I left class a little earlier. Everyone would have likely congratulated me [for making a good save]. But this is not what happened." The director is so conscious of *forking paths*, to use Jorge Luis Borges's expression,[57] that he alternates rigorously Otto's and Ana's versions of the facts. But even within the point of view of a single character, there are times when two versions of the same fact are made available. It is as if the following words of Borges had been given in a cinematographic form: "In all fictions, every time a man confronts different alternatives, he opts for one and eliminates the others; in that of the almost inextricable Ts'ui Pên, he opts—simultaneously—for all of them. In this way, he creates diverse destinies, different times, that proliferate and diverge."[58]

One day, after having fallen in love with Ana, Otto leaves the school. His father calls him from inside of an otherwise empty car. But Otto closes his eyes.

When he opens them, he finds himself still at the exit of the school. His father calls him again. But this time the car is no longer empty! Besides his father, Ana and his mother are also inside of the car.

Later on, when the adolescent Otto goes inside Ana's bedroom and cannot awake her, he returns to his own room and there is Ana waiting for him! After the death of Otto's mother, Otto and Ana are traveling through the snow on a sled. They are going at a high speed toward a precipice. Ana manages to get off the sled, but Otto does not and crashes. In the first version of this incident, Otto has died and an "angel-skier" rescues him and takes him to the summit. In another version, Ana looks for him and finds him alive. But as Ana searches for Otto, two other parallel actions emerge. In the first, the sled is stuck within some trees when, all of a sudden, it falls on Ana and kills her. In the second version, the sled falls a few meters away from Ana and thus she comes out unscathed. The conclusion also branches into two possibilities. In the first, Anna and Otto encounter each other once again; the camera shows Ana's eyes reflecting the image of Otto as he gets closer to hug her. In the second, Otto runs toward Ana after having contemplated how she had been ran over by a bus; the camera once again shows the image of Otto reflected in Ana's pupils.

In sum, "nonseparability" connects us to everything that exists—and not simply in a metaphorical way, but rather in a way that is experimentally verifiable—from the closest to the most remote and far away. This sort of connection between subjectivity and objectivity, where everything is reciprocally influenced, has been expressed artistically beginning with expressionism. The only difference between expressionism and quantum aesthetics lies in the fact that the artists and writers of the German movement, perhaps as a result of having predicted the slaughter of World War II, felt estranged and detached from the world despite having colored it with subjectivity.[59] In contrast, quantum writers and artists understand the extent to which they are actually interrelated with the world. Quantum artists are not stranged by the world, but rather understand its meaning. In fact, they—as well as everybody—are the sources of such meaning; they are the mediators of everything that exists and occurs. This message is conveyed in Francisco Plata's poem "La Jaula" ("The Cage"), in which the caging of a dog and the responsibility for the pain that is inflicted on this animal resonate throughout the universe:

> He (the dog) is nothing more than a corner
> of a glacial and pestilent warehouse
> situated in a minute town
> in the southern region of a modern country,
> one of the many states that has engendered
> this planet, third in a system
> that revolves around a distant star
> in the center of a galaxy that moves millions
> within its spirals, who knows how many;
> a galaxy that is rather small on this side of the universe,

of this universe that resonates
the cries of that puppy that you and I have caged.[60]

Aesthetically, we can also conceive that whatever happens in holographic sequences summarizes the before and after, as if everyone's life comprised a few critical points where everything is contained—something akin to Borges's *alephs*.[61] This is what takes place in Fernando de Villena's *El Fantasma de la Academia* (*The Ghost of the Academy*),[62] where a series of nuclei or black holes distill time.

INTUITION, TELEPATHY, AND TIME TRAVEL

Aspect's experiment, and its conclusion that the universe is not local, bolsters the existence of phenomena such as intuition and telepathy. For the first time, hunches, which have driven great literary characters to their discoveries, are not merely literary resources, but rather essential truths that we have to rely on and learn how to use. On the one hand, we can intuit the entire universe, its form, laws, and mysteries, and these, on the other hand, have a decisive influence on us. The universe unfolds as our dominion, without us having to leave our seats. This has never been addressed seriously in art or literature. For this reason, the painter Xeverio can say that, with his *petrale*s, and without needing a space ship, he is investigating the limits of the cosmos and of matter.

Subsequent to Aspect's experiment, the idea of time travel became a possibility. Putting science fiction aside, which has repeatedly entertained this theme, the possibility of time travel has produced today some disquieting accounts that redefine time while subverting essential truths. This is what takes place in Borges's "El Otro" ("The Other"), where the mature writer of 1969 meets with the young Borges of 1918 in a bank in Cambridge/Geneva:

> "In that case," I said straight out, "your name is Jorge Luis Borges. I, too, am Jorge Luis Borges. This is 1969 and we are in the city of Cambridge."
> "No," he said in a voice that was mine but a bit removed. He paused, then became insistent. "I'm here in Geneva, on a bench, a few steps from the Rhone."[63]

Similarly, in Julio Medem's *Los Amantes del Círculo Polar*, the protagonists, especially in moments of great intensity, are afforded the ability of being the children that they once were, as if time was ordered by content rather than by events.

As this new *multidimensional* man explores his own depths and attempts to grasp the infinite domain where he resides, art and literature play a crucial role, perhaps even more important than that played by "naked science." Art and literature cease to be mere artifacts of pleasure, and instead become necessary instruments of knowledge. That is to say, art and literature become vital tools for understanding the human psyche and take as their objective the task of mapping a world that, no matter how often we exploit its treasures, can never be extinguished. Art and literature have thus become something absolutely necessary in the study of knowledge, indispensable for the work of the "quantum man," who is expected to achieve self-knowledge, or, stated differently, to "individuate."

INDIVIDUATION

"Individuation" is the psychical equivalent of the complexity principle that regulates the universe. Just as small structures tend to group themselves into larger, more complex units, humans should learn to distinguish life's multiple components in order to understand its landscapes, labyrinths, and inhabitants. The integration of these multiple components, moreover, must proceed in such a way that avoids superimpositions (i.e., where one component dominates others).

Quantum aesthetics contributes to individuation. Quantum art and literature do more than reveal the most diverse constituents of the psyche, promote harmony among them, and illustrate in words and forms how to accomplish these aims. Additionally, all quantum manifestations are in tune with the universe and move toward complexity and self-knowledge. Quantum aesthetics, accordingly, encourages this trend.

One of the first novels that is consistent with this sort of focus is Miguel de Cervantes's *Don Quixote*. But how can this be possible when this novel was written four centuries ago? One of the traits of literary masterpieces is that they synthesize truths that are yet to be discovered. *Don Quixote* is both a modern and postmodern novel, and may even embody many other aesthetic forms that are not yet known. In this work, Cervantes lays out the history of Alonso Quijano's individuation. Quijano is a man who was deeply immersed in books and cherished the comfort of his home, but had forgotten about his own inner growth. So what happens? He goes insane. He goes insane because the pressure to individuate is so strong that, when everything else fails, insanity becomes his last resort. Insanity, therefore, might be what drives Don Quijote through his own inner ghosts, worlds, and personalities toward his final objective: gaining self-knowledge.

We are not singular but rather a multitude of beings (or psychic energies). This discovery is nothing but a psychological version of the principle of complementarity. In all men there exists an *animus* (*anima* in women), a *shadow*, a *sage*, a *scoffer*, a *twin*. Pauli had no reservations about comparing all these parts to the diverse elements of an atom. Art, like physics, can be a medium to expose these components and literature a means to know them.

This is precisely what Cervantes does. The exterior world where Don Quixote lives is a copy of his own inner world. Cervantes shows the quantum characteristic of exploring the invisible without departing from visible reality. This is why he has been naively called a realist. But what Cervantes explores through the landscape of la Mancha is Don Quijote's psyche. In the same manner, Sancho, his companion, is of flesh and bone, but he also represents the *shadow* of his master. That is, Sancho is Don Quixote's reversal and contains all that his one-dimensional personality has thrown overboard. Accordingly, as the novel unfolds, Don Quijote and Sancho, like a proton and a neuron that unite to form a nucleus, are gradually integrated, and Don Quijote "sanchifizes" as Sancho "quijotizes." We should do nothing else with our shadow and all the other beings that inhabit us!

Similarly, fuzzy thinking confirms that we can simultaneously embody the most contradictory of personalities. Because of this *complexity* or fuzziness in their personalities—complementarity of impulses and behaviors—William Shakespeare's characters—like those of Cervantes—are entirely human and their lives often escape the literary work where they reside. In the end, Don Quijote achieves wisdom because all his introspective work results in self-knowledge, maturity, and complexity. At the end of his trip, Alonso Quijano finally becomes himself; that is to say, he has become *individuated*.

SYNCHRONICITIES

In Plata's poem reproduced earlier, a causal relation with the universe takes place. However, this relation can also be acausal; that is to say, this relation is not necessarily the outcome of a particular or single action. This leads us directly to the concept of synchronicity or "significant coincidence." Because mind and matter appear to be derived from a common source, it is possible that they have respective manifestations that are in sync with one another. Implied is that a thought can be materially *rejected, confirmed*, or *contested*. A well-known synchronicity narrated by Jung took place when a patient referred to his dream about an Egyptian scarab. In that moment, something hit against the window. Jung opened the window, and noticed that in front of him, dizzy from the impact, was a scarab![64] Ann Ulanow, moreover, narrates the following synchronicity: "[A] patient battled against his childhood fear of being locked inside an attic that had served as his punishment for protesting against being sent to sleep at night. He was finally able to discover the key to a compulsive fetish that had served as the symbol that linked his adult personality with his childhood fear of being locked inside that dark place. When this new attitude emerged in the midst of his battle between his fetish fascination on the one hand, and his conscious humiliation and desire to get rid of his compulsion on the other, an external synchronicity was produced. The small dark room was destroyed by a lightning flash that struck only that part of the house!"[65]

Synchronicities have been taken into account intuitively by writers throughout history, and, most notably, in gothic and romantic literature. These two genres of literature are full of encounters, nonencounters (*desencuentros*), significant accidents (*azares*), and character recognitions (*anagnórisis*), although all of these have been viewed as suspicious, as if the writer has a license to fantasize, to the point of being relegated to inferior modes of expression by twentieth-century realism. Nonetheless, we now face the discomforting fact that synchronicities actually occur. We must consider, however, that not all coincidences are synchronicities. For the latter to occur, the former must be significant, that is, there has to be a connection between the event and somebody's thoughts.

Some writers and artists have begun to use the notion of synchronicity in their works with disturbing results for both readers and viewers. This is the case of movie director Krzysztof Kieslowski, who uses the notion of synchronicity in his trilogy *Blue*, *White*, and *Red*.[66] To give a few examples: *Red* ends with the shipwreck of a ferry in the Straight of Dover. Of fifteen hundred passengers, only seven survive. With the exception of the ship's bartender, who is the only survivor about whom we know nothing, the rest of the characters' lives are unconsciously interrelated, via the shipwreck—a synchronicity. Of course the viewer who has watched and given some thought to the previous movies may discern three couples among the survivors: Olivier and Julie, the protagonists of *Blue*; Karol and Dominique, the protagonists of *White*, and Maurice and Valentina, the protagonists of *Red*. All three couples have been crossing paths time and again throughout the film, without ever directly encountering one another. Synchronicity is what unites each of their destinies to that of the other couples. Stated otherwise, Kieslowski's trilogy is a *flashback* of the lives of these six protagonists that the shipwreck has united. In other words, the entire body of work reveals a synchronicity that unites the protagonists' individual destinies into a collective one.

But there are many more synchronicities in Kieslowski's trilogy that are developed through a wide range of real and contemporary gradations and possibilities, which, at the same time, are united with the universe and the cosmos. To give another example from *Red*, the news of the shipwreck is being televised when a photograph of Valentina suddenly appears on the television screen. The photograph is zoomed in on and turns out to be the same one that Valentina had posed for in an advertisement the month earlier. This is the same picture that has been shown on numerous billboards throughout the city and that, in turn, begins to be removed at the moment that the storm that causes the shipwreck has formed. Valentina's advertisement photograph, in other words, is a precognition, a prophecy, a *real* image of the future. That is, in Kieslowski's opinion, the future can actually be read and can even penetrate the present. He portrays the mystery of life so well that the viewer feels continuously the presence of this mystery and begins to believe that it will be revealed at any moment.

Synchronicities also appear in *Los Amantes del Círculo Polar* to such an extent that what the characters say about *significant coincidences* (they simply call them coincidences) can actually comprise a textbook on the issue. When Ana finally settles in a Finnish cabin that is situated in the polar circle, she states: "I will remain here for as long as it takes. I am waiting for the coincidence in my life, the biggest one of them all. This is the case despite the fact that I have experienced all kinds of coincidences. Yes, my life can be told as a series of coincidences." We can also find numerous synchronicities in Auster's *Leviathan*.

ARCHETYPES

Sheldrake's morphogenetic fields can be translated psychically into Jung's concept of archetypes, which refer to the forms, rules, and behaviors that are manifested in nature and people.[67] Archetypes, much like morphogenic fields, are ever-changing. Making an explicit comparison between archetypes and the subatomic particles that are formed by corpuscles and waves, Jung states that "an archetype (as a structural element of the unconscious) is composed, on the one hand, by a static part, and, on the other hand, by something dynamic."[68] An archetype is static because it persists throughout time as a pattern of nature. At the same time, as every historical epoch "realizes" this pattern, the archetype undergoes transformations. For example, the archetype of the hero who confronts a dragon is as recognizable today as it was four thousand years ago. But four centuries ago, Cervantes transformed this archetype into the form of a mad man who battles windmills; and, one hundred years ago, Henrik Ibsen transformed the same archetype into the form of a dignified man who battles generalized and corrupt opinions.[69] And today, the same archetype is being transformed continuously by a host of Hollywood movies with actors such as Arnold Schwarzenegger.

Art and literature, therefore, play an important role in the ongoing transmutation of archetypes and their historical adaptation. Thus, the more a literary or artistic work embodies an archetype, the more it is seen as beautiful, moving, and fascinating. All the world's great literary works embody archetypes; that is, they all connect with some of the most profound facets of the collective unconsciousness. According to Jung, real art sheds light on the most necessary archetypes of its respective historical period. As a result, artists do not simply confirm the mentality of their time, but rather contradict, suspend, and fracture this outlook. Typically, their contemporaries find this sort of practice alienating. However, "the relative maladaptation of the artists is also their true advantage," because "this allows them to be isolated from the general currents, yield to their own desires, and find what others unknowingly lack."[70] One of the characteristics of quantum artists is that they seek that which is not visible, that which underlies something else, that which is ignored; that is to say, in the face of the *explicate order*, the quantum artist looks for the *implicate order*. In contrast to the obtuse residents of the small house, quantum creators are like visionaries who contemplate what normal eyes cannot see. Their objective is to open a door "through which one can penetrate, on occasions from a non-human world, the unknown, that which acts secretly."[71]

For Circlot, art opens a window to a concealed world that we can enter because of our imagination. In reality, science is sustained by fantasy. Without the human imagination, scientific theories would not exist, and much less subatomic theories. For this reason, Circlot believes that "imagination is the most scientific of all faculties."[72] In fact, Circlot goes so far as to identify the imagination as being inextricably connected to reality. In his opinion, the former can at times be more real than the latter. Rebelling against the idea that reality surpasses both art and literature, he states: " 'Imaginary sentiments' are not inferior to normal sentiments, but rather superior."[73] Circlot has no problem in concurring with Novalis that "we are more intimately connected to the invisible than the visible," and in this regard comments that "[i]f one of his [man's] domains is the surface of the planet, the other consists of the depths of his life."[74]

Within these depths lies the mystery that Circlot seeks to unravel in his work as a poet. Thus, he has no doubts in asserting that "it is easier to sympathize with those, like Poe and Nerval, who launched themselves into the sea of the imaginary, 'of the forever lost.' "[75] These are the types of authors that Circlot seeks and finds interesting; authors who see reality "as a much larger domain than what most writers and psychologists usually admit."[76] Of course, when he writes about authors and psychologists, he is referring to those of his time—the article appeared in 1966—whose understanding of aesthetics is still grounded on social realism, including empty experimentation or Freudian psychology. According to Circlot, the authors who *saw reality in a much more profound way* included, other than Edgar Allen Poe, Gérard de Nerval, William Blake, William Shakespeare, H. P. Lovecraft, George Trakl, Alexander Blok, André Breton and James Joyce in literature; Igor Stravinsky and Arnold Schoenberg in music; and Joan Miró, Antoni Tapies, Modest Cuixart, and Pablo Picasso in painting.

In sum, Circlot seeks access to the "fuzzy" world inhabited by undiscovered elements. As quantum physicists recognize, reality can never be known directly. According to Circlot, this is the task of symbols. As a result, he asserts that "symbolic artists of the 20th century spoke about what did not have a name."[77] Indeed, symbols are the instruments that allow us to apprehend the unknown. The importance that Circlot gave to symbols is manifested in the fact that he paid a great deal of attention to dreams, as well as to the history of symbolism that he presents in his masterful *Diccionario de Símbolos* (*Dictionary of Symbols*).[78]

In general, all forms of realism have neglected the hidden patterns of archetypes. Only a handful of insightful authors have perceived that "unconsciousness not only influences consciousness, but rather guides it."[79] One such author is Henry James, whose characters see that their conscious volitions are constantly betrayed by something that overwhelms them, and that the structures that lead to this betrayal become manifested to the readers only through the characters' actions. But beyond these clairvoyant authors, quantum aesthetics illustrates that the differences among the models that the characters manifest (or that art works embody) are not the result of the creators' mere intuition, but rather their response to a scientific and well-founded view of life. That is, quantum authors deliberately seek specific archetypes in their work, and they illustrate them by intervening in their characters' lives, much like the archetypes would do in their own real lives.

There are also writers who return directly to myth, as is the case of the composer Lawrence Axelrod. In his work *Cassandra Speaks*,[80] the composer revives the myth of the Trojan heroine who has not only been given the gift of prophecy, but also the burden of not being believed by others. The work ends in gradual silence: recognizing the futility of obtaining real knowledge in a mendacious world, we return to a total understanding of quantum nothingness.

Another example is found, again, in Medem's *The Lovers of the Polar Circle*, where archetypes play a very important role. For example, the director creates different geometric configurations in the film, starting with the names of the protagonist couple (Ana and Otto), which are symmetrical and, as a result, can be reversed. Many more symmetries follow. During the bombardment at Guernica, a Spaniard (Álvaro) frees a German aviator's (Otto) parachute from some tree branches and they become friends. As a sign of gratitude, Otto names his German son Álvaro, while Álvaro names his Spanish grandson Otto. Otto the grandfather had married a Spanish woman, while Álvaro's son eventually marries a German.

More symmetries: The manner in which Otto the grandfather meets his future wife is identical to the way that, many years later, Ana and Otto meet as children. After abandoning his airplane and parachuting down, Otto the grandfather gets stuck among some tree branches; the same thing happens to Otto the son after having abandoned the postal aircraft he piloted between Spain and Finland. The geographic term ("polar circle") that appears in the movie's title is also a geometric allusion, for the term symbolizes the circles that comprise our lives. The movie shows us the historical development of the first circle that encompasses Ana and Otto, which begins when they are eight and ends when they are twenty-five.

On other occasions, archetypes are manifested through the repetition of specific motives or symbols. One of these symbols that is iterated throughout the movie is the airplane. Otto the grandfather is an aviator; Otto the son makes paper airplanes; when Ana is a teacher, she becomes fascinated by a paper airplane that her students have made; and Otto the son also becomes a pilot.

The outlining of archetypical figures, or *mandalas*, is also one of the procedures employed by Antonio Arellanes in his paintings. Refering to Arellanes, Leonard Shlain states:

> His iconography presents archetypical images juxtaposed in such a fashion that at some unconscious level we are challenged to question our assumption concerning these three basic constructions of reality.... Antonio inserts symbols into his complex compositions that are windows into other levels of conciousness.... He also imbues these symbols with mythic language that connects them to the Collective Unconscious; that reservoir of half-seen symbols that portray the *imago mundi*, the image of the world as seen through the filters of our limited perceptual apparatus.... By mixing images from ancient Native traditions with those that evoke machinery, the artist compresses both the time and space of the human condition and presents in one canvas ideas that undergird both the modern and the archaic.[81]

BEAUTY

We have already seen how beauty, for scientists, constitutes the most convincing proof for the accuracy of their theories. Although beauty has been traditionally understood as intrinsic to art and literature, this connection was cast aside during the twentieth century for the purpose of revindicating—either implicitly or explicitly—ugliness. In this respect, artistic and literary expression broke from ideals of beauty. Art became disengaged, divided, and detached—cubism is a good example. Literature shred syntax to pieces, rejected clarity, and disturbed harmony. All of this evolved into the vertiginous and *inhumane*[82] telegraph phrases of the futurists, or the disembedded, unmediated consciousness of Joyce's *Ulysses*. Realisms, however, arrive at eschatological grounds and replace fluids for emotions, defecation for thought, and society for *detritus*. Of course, all this is for a reason: the incorporation of what art had for many centuries marginalized in the name of "good taste." Accordingly, the need to express this repressed shadow became a requisite for mental sanity, and this task was assumed during the twentieth century, although, as is typically the case, it was taken to an extreme. All the while, scientists, much like Medieval monks, protectively possessed the key to beauty.

Today, however, it has become absolutely necessary to reintegrate beauty into art and literature. Yet, the idea is not to reproduce an old conception of beauty. If the world were comprised solely of light and beautiful forms, it would be something *rococo* or *pompier*. The new concept of beauty emerges out of its integration with ugliness. Suffering, destruction, and horror cannot be simply discarded; instead, they should be understood as creative dimensions that form a part of a totality. Real beauty emerges when the *Apollonian*—beauty as used to be defined—is integrated with the *Dionysian*—all that is bad, horrendous, dissolute, and destructive. From a dialectical point of view, we could say that a *thesis* (beauty as pure luminosity and beautiful forms) took root during the eighteenth and part of the nineteenth centuries (with movements such as Parnasianism), while the *antithesis* also flourished during part of the nineteenth (romanticism and symbolism) and all of the twentieth centuries. Finally, the *synthesis* that quantum aesthetics establishes will undoubtedly inform a great part of the twenty-first century.

The works that have been cited thus far illustrate key quantum aspects that are imbued within this understanding of beauty. This is the case in Kieslowski's trilogy, which starts with a shipwreck in which fifteen hundred people perish, or in *Lovers of the Polar Circle*, where the characters are constantly surrounded by death. Another example is Auster's *Leviathan*, in which the story begins to unfold after Sachs is killed in an explosion. This quantum understanding of beauty is also found in Xaverio's paintings, where we find bright, cheery colors juxtaposed with darker, more somber tones. We also find this is Plata's poem, in which he shows the suffering of an animal echoing throughout the universe; in the work of Arellanes, who alternates Apollonian and Dionysian symbols; in Villena's work *The Ghost of the Academy*, where darkness, frailty, and decomposition constantly confront light, form, and volition; or in the writings of Circlot who sees beauty in death[83] and pain.[84] In the end, we can conclude with some qualification that works of art or literature are not quantum if they are not beautiful, while understanding beauty as a *coiunctio oppositorum*, that is, an integration of light and shadow, form and amorphism, plenitude and nothingness.

OTHER ARTISTIC CONSEQUENCES

So far, we have discussed a few of the artistic possibilities that are derived from quantum principles. But there are many more. For example, by identifying language as visible matter, we can decompose it as a way to explore its more profound elements. This would lead us to understand (1) *language* and *text*—without differentiating one from the other—as being tantamount to visible matter or raw materials that people mold according to their own interests and knowledge, (2) *language* as being a subjective vision of the world that is shared by a group of speakers, and (3) *metalanguage* as the study of the ideological components of a concrete text. We can say that *language* is the instrument used to observe an evident reality (embedded within language or text), whereas *metalanguage* uncovers the internal and hidden laws of this reality.[85] Miguel Ángel Diéguez understands these distinctions. In all his novels, he questions the notion of a *text* as an *objective reality*, breaking reality into multiple parts, and, as a result, unmasking language in such a way that narratives are converted into the very histories whereby their texts have been elaborated or generated. Accordingly, Diéguez unravels the evident reality of a text and flagrantly denounces its *ideograms*,[86] that is to say, the ideological and mental categories that a text conceals. This is certainly the case in his work *En la Gran Manzana* (*In the Big Apple*), which begins like this: "I slid into the chain of contiguous rooms in the Big Apple. The crowd appeared to be surrounded by a halo created by the extraordinary lights in the false roof. The music and whispers entwined themselves with the depths of my mind. All of the creatures searched themselves, anxious to get a fix of topical language."[87]

There are still many more possibilities. Of course, it would be impossible to offer an exhaustive list. First, space is limited in this chapter. And second, because these possibilities are affected by the future creativity of artists, and thus we cannot foresee what new forms will emerge. In any case, we should conclude by agreeing with Davies that reality, and, by implication, literature and art, are more similar to Lewis Carroll's *Alice in Wonderland* than Balzac's *Human Comedy*.[88] What we have in quantum aesthetics is a full blown revolution.

THE QUANTUM AESTHETICS GROUP

Although it is difficult to decipher the origins of something because everything is intertwined and can be traced from precedent to precedent, the most immediate origins of the Quantum Aesthetics Group are found in the Salón de Independientes (Independents Hall), where a group of writers rendezvoused in 1994 with the objective of defending creative freedom from political or other interferences. At the heart of this movement was the creation of a new aesthetic that opposed official canons and advocated a resurrection of the archetypes that are most necessary for the present epoch. Accordingly, as is the case with everything that is new, the movement swam against the tide. Quantum aesthetics germinated out of a series of discussions among the *independientes* (independents). One of its first explicit formulations was advanced at a lecture entitled "La Diferencia Cuántica" ("The Quantum Difference") that took place in Valencia, on June 9, 1995, at the conference La Diferencia Posible (The Possible Difference).[89] Subsequent to this conference, a series of discussions, other conferences, and monographs followed, many of which are discussed in my book *El Cadáver de Balzac*.[90] It was then that the idea of forming a group united by the objective to develop a new aesthetic was considered, and this prospect came to fruition in 1998. Although the group was created with the intention of incorporating all types of creators, its initial base consisted of the writers Miguel Ángel Diéguez, Francisco Plata, Ángel Contreras, Fernando de Villena, Rosario de Goróstegui, Arturo Zamudio, and myself. Soon after, we had to opportunity to meet the painter Xaverio, who, in his work, had come to conclusions similar to ours. When he joined the group, a host of other plastic artists followed, such as María Caro, Andrés Monteagudo, Agustín Ruiz de Amodóvar, Joan Nicolau, and others.

In a meeting that took place in February 1999, a manifesto (or *E-m@ilfesto* as it was later called, because it was disseminated via the Internet) was created. In this document, the point is made that the group's primary objective is to "to explore the creative avenues open by the most advanced science and psychology, turning art and literature into a precise knowledge tool capable of inquiring about human complexity and everything that surrounds humankind."[91]

The group later began to grow and became international with the incorporation of German photographers Sussana Majorck and Thomas Busse; the Argentinian writer José Gabriel Ceballos; the anthropologist, also Argentinian, Graciela Elizabeth Bergallo; the Spanish psychologist Juan Antonio Díaz de Rada; the North American musician Lawrence Axelrod; the North American painters K. C. Tebutt (Canada) and Antonio Arellanes (United States); the Belgian Web designer Luc Schokkelé; and the hispanists of diverse nationalities Jennifer Wilson, Mihaela Dvorac, Francisco Peñas-Bermejo, Coman Lupu, and Andrea Vladescu.

On April 15, 1999, the group was officially introduced at the Galería Contemporánea, Granada, Spain, where Xaverio's collection entitled "Estética Cuántica" ("Quantum Aesthetics") was being exhibited. From that moment on, a myriad of encounters and meetings took place such as the "quantum dinner" celebrated at the farm of Sussana Majork and Thomas Busse in May 1999, while, on the Internet, various discussions among members from all parts of the world were also taking place.

A significant moment occurred when Manuel J. Caro, an instructor and Ph.D. candidate in sociology at the University of Miami, joined the group. Being familiar with quantum aesthetics, Caro instigated a fruitful dialogue between John W. Murphy, a professor in the Department of Sociology, and myself. As editors of the present volume, both are devoted to the project of developing a quantum culture.

Of course, the extension of ideas that necessarily signify what is to come, and, as a result, are in direct opposition to mainstream thought have not been spared from insults and stinging criticisms. In the beginning, when I began to give my first lectures, the *residents of the small house* felt a dark and yet unclear threat, and their reactions often took the form of pitiful irony and ridicule. One reporter who attended my conference "La Transgresión del Camino Literario Cuántico" ("The Transgression Embodied by the Quantum Literary Path") wrote the next day that it was a shame that my intervention had not been accompanied by "test and assay tubes, stills, and crucibles."[92] Another reporter criticized me for defending Don Quixote and denounced me as a rightist. In the newspapers, there appeared letters from indignant readers, one of whom accused me of promoting with quantum aesthetics not only "epistemological barbarism," but also the "very negation of epistemology." In an eight-page anonymous letter sent to the Valencian newspaper *Las Provincias*, I was accused, among other things, of being an "integrated circuit" writer, "a collaborator who is armed to the teeth," proposing "methods and strategies inspired by Nazism," and even advocating "genocide." And, to make matters worse, I was encouraged to "urgently seek the services of a good psychiatrist."[93] The ambiance, in short, was so blatantly scornful and hostile that when Wilson asked a guru of one of the old paradigms what he thought about quantum aesthetics, he responded "he was only a poet" and disdainfully turned his back to her.[94]

Later, in an e-mail debate with professors and students at the University of Oviedo, one of them discussed with scandalous irony the idea behind Schrödinger's thesis that man creates his own reality as follows: "Can we consider that the chimpanzees of the Great Ape Project,[95] who are capable of sweeping and dusting, are also cultural animals and 'creators' of their own reality? Perhaps they can even be listed among the names on the E-m@ilfesto, who knows." Another individual accused our ideas of resulting in "vagarious links between subatomic physics and aesthetic categories," or in "a combination of erratic displays and slogans that are illogical and ultimately devoid of any sense."

But all new ideas are contagious and their vigor foils silence, disdain, or any other effort to neutralize them. Despite all resistance, quantum aesthetics kept expanding, and, in the face of real danger, the *old order* decided to appropriate some of its ideas, albeit the most simple and obvious ones. For example, Wilson states that, one year later, when she returned to Spain, she found an article in the Spanish newspaper *El País* by the same guru mentioned earlier; this time he was praising the "inevitable brotherhood" between the "sciences and the humanities," affirming that they are "two ways of establishing time."[96] But was he not only a poet?

A TOTAL PARADIGM

The variety of persons, professions, and nationalities that are now part of the group must be understood as an example of the enormous changes and research possibilities that quantum aesthetics can offer not only to art and literature, but also to various disciplines such as linguistics, philosophy, sociology, and even politics, some of which will be analyzed in-depth in the present volume. To cite a theme that is common to all of these disciplines, it is necessary to understand that quantum aesthetics is, above all, a way to promote freedom, open interpretation, and the study of complex systems. Thus, for example, quantum aesthetics and democracy are tied necessarily. Once the destiny of every man and woman is understood as unlimited and full of possibilities, any violence or arbitrary imposition that may hinder this development is viewed as unjust and even criminal.

Quantum principles are even being applied to the law. In his work *Validez y Vigencia* (*Validity and Currency*),[97] José Luis Serrano affirms that the validity of a norm depends on "an aesthetic trial for coherence." That is, the law begins to have traits similar to those identified by quantum physicists, who believe that the beauty of a theory is the most convincing evidence of its certainty. Or, similar to a mathematical equation, "the validity and currency of a norm depend on factors that are internal to the juridical system." For Serrano, on the one hand, there is a Newtonian norm that is understood to be unquestionable, ahistorical, analytic, mechanistic, and atomistic (in the sense that this norm is considered to be absolute and indivisible). On the other hand, there is a quantum norm characterized by the importance of the observer or interpreter—he makes all norms relative. What is important for the latter is not something autonomous, but rather an embodied structure that is established with the entire normative *corpus*. In this manner, "the norm is not an a priori, but rather is constructed according to each case and each interpreter. . . . The norm is not an element, but rather a relation; it is not an atom, but rather a quantum." And taking this relationship to subatomic physics even further, Serrano states: "[N]orms are not indivisible units, previously understood as conventions, but rather are *relational and probabilistic constructions, more similar to quantums of energy than atoms.*"[98] This, in sum, makes the new juridical science advocated by Serrano "complex, systematic, quantum, and holistic." Could it be any clearer?

Given the examples offered earlier, it does not seem illogical to suggest that quantum culture will progressively saturate all disciplines until it becomes a total paradigm. Undoubtedly, in the opinion of an increasing majority, this paradigm will inform, at minimum, the twenty-first century. We find ourselves before a new way of thinking, as original as the first Neolithic practices that, in a world of nomads, took thousands of years to extend throughout the planet—and they yet have not done so entirely. Similarly, this new way of thinking will progressively expand throughout the world, although, we believe, much quicker because of the advances in communication that have been made in the modern world. Nonetheless, quantum culture will require no less than a century to reach all corners of the globe; this is the time that a historical period as interconnected as ours requires in order to abandon its habits and prejudices. The twenty-first century will be the era of the struggle and extension of quantum thought, even if we name these ideas—capable of accounting for both the macro and the micro—"superstring theory" or anything else. In all of this, aesthetics, the queen of all disciplines (for aesthetics is synonym with internal coherence, that is to say, *beauty*) will play the crucial role in exposing the mysteries that are concealed below the surface of everyday affairs.

NOTES

1. Translated into English by Luigi Esposito.

2. Friedrich Schiller, *Naive and Sentimental Poetry and on the Sublime: Two Essays*. (New York: Ungar, 1966). Schiller opposes ingenuity to *sentimentality*. The latter is a characteristic of human beings who internalize and assimilate the object. Both distinctions correspond to what is more commonly known as *extroversion* and *introversion*.

3. Polarization is understood as the spatial orientation taken by a light wave as the wave moves away from its source.

4. John Gliedman, "Interview with Brian Joseph," *Omni* 4, no. 10 (July 1982): 88.

5. This experiment was once again confirmed, with much more advanced means, in July 1997 by a group of physicists at the University of Geneva directed by Nicolás Gisin. Making this city the center of the experiment, two photons were sent through a fiber optic cable in opposite directions: one, to the city of Bernex, 7.3 kilometers away, and the other, to the city of Bellevue, 4.4 kilometers from Geneva. At the end of each cable, there was a mirror. The idea is that the photons would either penetrate or be reflected in the mirrors. At all times, both photons behaved in the same manner, either running through the mirror simultaneously or being reflected in the mirror. See R. Ikonocoff, "Espacio y Tiempo: El Experimento que Desafía a Einstein," *Ciencia y Vida* 1 (March 1999): 72–76. More recently, several researchers from the Princeton's Research Institute have produced a beam that travels faster than light. See NEC Research Institute, "Gain-Assisted Superluminal Light Propagation," *Nature* 406 (July 2000): 277–279.

6. "Castilla miserable, ayer dominadora, envuelta en sus andrajos desprecia cuanto ignora." In "A Orillas del Duero," *Campos de Castilla (1907–1917), 9th ed. (Madrid: Cátedra, 1999)*.

7. Rupert Sheldrake, *Seven Experiments That Could Change the World* (London: Fourth Estate Limited, 1994).

8. In Paul Davies, *El Espacio y el Universo Cuántico* (Barcelona: Salvat Editores, 1994), 44. English version: Paul Davies, *Other Worlds: Space, Superspace, and the Quantum Universe* (New York: Viking Penguin, 1997).

9. See Edgar Morin, *Introducción al Pensamiento Complejo* (Gedisa: Barcelona, 1995).

10. Sheldrake, *Seven Experiments that Could Change the World*.
11. See Bart Kosko, *Fuzzy Thinking: The New Sciences of Fuzzy Logic* (New York: Hyperion, 1993).
12. For example, see Lofti Zadeh, ed., *Fuzzy Sets and Their Applications to Cognitive and Decision Processes* (New York: Academic Press, 1975).
13. Brian Greene, *The Elegant Universe: Superstrings, Hidden Dimensions, and the Quest for the Ultimate Theory* (New York: Norton, 1999).
14. David Bohm, *Wholeness and the Implicate Order* (London: Routledge and Keagan Paul, 1980).
15. In Ken Wilber, ed., *Cuestiones Cuánticas* (Barcelona, Kairós,: 1987). English version: Ken Wilber, eds., *Quantum Questions: Mystical Writings of the World's Great Physicists* (New York: Shambhala, 1985).
16. Carl A. Meier, ed., *Wolfgang Pauli y Carl G. Jung: Un Intercambio Epistolar, 1932–1958* (Madrid: Alianza Editorial, 1996), 153. English version: Carl A. Meier, ed., Wolfgang Pauli and Carl G. Jung: *Correspondance, 1932–1958* (Paris: A. Michel, 2000).
17. See Elisabeth K. Valkeinier, *Russian Realist Art: The State and Society: The Perdzvihniki and Their Tradition* (New York: Columbia University Press).
18. Hundreds of Russian writers were kept in this prison, which is situated in the center of Moscow. Any dissent from official impositions was enough to be put in there. Many writers incarcerated there, such as Isaak Bábel, were executed in the prison's basement. This sort of thing was so common that the names of all of those who were executed were squeezed, with very little space between them, in hundreds of archival books. However, it appears that during the Soviet period, fifteen hundred writers died in concentration camps. See Vitali Chentalinsky, *De los Archivos Literarios del KGB* (Barcelona: Anaya and Mario Muchnik, 1994).
19. For example, see Bert Schierbeek, *The Experimentalists* (Amsterdam: J. M. Menlenhoff, 1964), 5–17.
20. John L. Ver Ward, *American Realist Painting, 1945–1980* (Ann Arbor: University of Michigan Research Press, 1989).
21. See Enrique Anderson Imbert, *El Realismo Mágico y otros Ensayos* (Caracas: Monte Ávila Editores, 1976), especially 7–25. See also Gloria Bautista Gutiérrez, *Realismo Mágico, Cosmos Latinoamericano: Teoría y Práctica* (Bogotá, Colombia: America Latina, 1991), especially 13–31.
22. Pablo Francescutti, "Un Siglo Cuántico: Entrevista al Físico Francisco J. Yunduráin," *El Cultural (Diario ABC)* 19 April, 2000, 61–63.
23. Fragment of a letter to Louise Colet, April 24, 1852. Gustav Flaubert, *Cartas a Louis Colet* (Madrid: Siruela, 1989), 183–184.
24. Leonard Shlain, *Art and Physics: Parallel Visions in Space, Time, and Light* (New York: Morrow, 1991), 427.
25. Edward Osborne Wilson, Consilience. The Unity of Knowledge. (New York: Knopf, 1998). Spanish version: Edward O. Wilson, *Consiliencia: La Unidad del Conocimiento* (Barcelona: Galaxia Gutenberg/Círculo de Lectores, 1999).
26. Wilson, *Consiliencia, 15*.
27. Ibid., 310.
28. For example, see the narratives compiled in José Gabriel Ceballos, *El Patrón del Chamamé* (Alvear, Corrientes, Argentina: Ediciones Río de los Pájaros, 1998).

29. ("ningún pensamiento humano es capaz de prever, en estos momentos, a qué resplandecientes soles psíquicos se abrirán las obras en el futuro.") Fragment of a letter to Louise Colet, April 24, 1852. See Flaubert, *Cartas a Louis Colet*.
30. "Spin" refers to the internal rotation of a particle.
31. Paul Davies, *Superfuerza* (Barcelona: Salvat Editores, 1994), 30–31. English version: Paul Davies, *Superforce: The Search for a Grand Unified Theory of Nature* (London: Penguin, 1995).
32. Davies, *Superfuerza*, 31.
33. Miguel Ángel Sabadell, "La Fisica Moderna Podría Ayudarnos a Entender la Conciencia," *Muy Interesante* (February 2000), 115–117.
34. For example, see Jorge Luis Borges, Epílogo to *Prosa Completa, vol. 2, Otras Inquisiciones* (Barcelona: Bruguera, 1980), 305.
35. See Juan Eduardo Cirlot, "El pensamiento de Gérard de Nerval," in *Confidencias Literarias* (Madrid: Huerga y Fierro Editores, 1996).
36. See Carl G. Jung, *El Hombre y sus Símbolos* (Barcelona: Caralt, 1976), 266, 331. English version: Carl G. Jung, *Man and his Symbols* (London: Aldus, 1972).
37. Marie-Luise von Franz, "La Ciencia y el Inconsciente," in *El Hombre y sus Símbolos*, by Carl G. Jung et al. (Barcelona: Caralt, 1976), 328.
38. Ibid., 331–332.
39. See Gregorio Morales, "El Ancla Oculta de la Estética Cuántica," in *El Cadáver de Balzac* (Alicante: Epígono, 1998).
40. Meier, *Wolfgang Pauli y Carl G. Jung*, 100.
41. Herbert Marcuse, *One-Dimensional Man* (Boston: Beacon, 1964).
42. Daniel Goleman, *Emotional Intelligence* (New York: Bantam, 1997).
43. Michael Frayn, *Copenhagen* (London: Methuen Drama, 1998).
44. James Glanz, "Of Physics, Friendship and Nazi Germany's Atomic Bomb Efforts," *New York Times*, 21 March 2000, emphasis added.
45. Ibid.
46. See Gregorio Morales, "Una Literatura Borrosa," in *El Cadáver de Balzac*, 53–58.
47. The "petrales" are combinations of crystals, gems, and minerals crushed into particles that are glued with pigmented residues over wooden sheets.
48. See Chapter 6 in this volume.
49. Paul Auster, *Leviathan* (New York: Viking Penguin, 1992).
50. Paul Auster, *Leviatán* (Barcelona: Anagrama, 1993), 139.
51. ("De no haber sido por la ruptura de mi matrimonio con Deli Bond, nunca habría conocido a Maria Turner, y si no hubiera conocido a Maria Turner, nunca me habría enterado de la existencia de Lillian Stern, y si no me hubiese enterado de la existencia de Lillian Stern, no estaría aquí sentado escribiendo este libro.") Auster, *Leviatán*, 65–66.
52. Juan Eduardo Cirlot, "Los Restos Negros," in *Obra Poética* (Madrid: Cátedra, 1981), 195–202.
53. Charles Baudelaire, "Correspondance," in *Les Fleurs du Mal* (France: Gallimard, 1972), 38. English version: Charles Baudelaire, *Fleurs du Mal* (London, Cassell, 1952).
54. ("¿Correspondencia? El arte de los monjes Irlandeses en la epoca anterior al período carolingio y durante éste, arte basado en intricadas lacerías, en entrelazamientos que constituyen laberintos indefinidos, absolutos, es la mejor prueba de que en "este

mundo"—y acaso sea ésta su definitiva justificación—todo se corresponde, enlaza y comunica.") Juan Eduardo Circlot, "El Retorno de Ofelia," in *Confidencias Literarias* (Madrid: Huerga y Fierro Editores, 1996).

55. (Todo está relacionado con todo, cada historia se solapa con las demás. Por muy horrible que me resulte decirlo, comprendo ahora que yo soy quien nos unió a todos. Tanto como el proprio Sachs, yo soy el punto donde comienza todo.") Auster, *Leviatán*, 66.

56. Julio Medem, *Los Amantes del Círculo Polar* (Sogetel, Spain, 1998).

57. This expression relates to Jorge Luis Borges, "El Jardín de los Senderos que se Bifurcan," *Prosa Completa*, vol. 1, Ficciones. Barcelona: Bruguera, 1980. In this story, the labyrinth is established as a symbol of the parallel universes and the other infinite possible universes that emerge from them. English version: Jorge Luis Borges, "The Garden of Forking Paths," *Labyrinths: Selected Stories and Other Writings* (Norfolk, CT: James Laughlin, 1976), pp. 19–29.

58. ("En todas las ficciones, cada vez que un hombre se enfrenta con diversas alternativas, opta por una y elimina las otras; en la del casi inextricable Ts'ui Pên, opta—simultáneamente—por todas. Crea, así, diversos porvenires, diversos tiempos, que también proliferan y se bifurcan.") Borges, "El Jardín de los Senderos que se Bifurcan," 369–379.

59. See Douglas Keller, "Expressionism and Rebellion," *The Expressionist Heritage*, ed. Stephen Eric Bronner and Douglas Keller (New York: Universe, 1983). See also Bernard S. Myers, *The German Expressionists* (New York: Praeger, 1957), 11–14.

60. See the original Spanish version in Francisco Plata, "La Jaula," *Calas* 5 (June 1999): 215.

61. Jorge Luis Borges, *The Aleph and other Stories, 1933–1969* (New York: Dutton, 1970).

62. Fernando de Villena, *El Fantasma de la Academia* (Granada: Port-Royal, 1999).

63. Jorge Luis Borges, "The Other." *The Book of Sand* (New York: Dutton, 1977), 12. This story was written before Alain Aspect's experiment (1982), although the paradox raised by Einsten, Podolki, and Rosen (1935) had already suggested the possibility of time travel.

64. See Carl G. Jung, *Memories, Dreams, Reflections* (New York: Random House, 1963).

65. Ann Ulanov, "Jung y la Religión: El Sí-mismo en Oposición," in *Introducción a Jung*, ed. Polly Young-Eisendrath and Terence Dawson (Madrid, Cambridge University Press, 1999), 426. English version: Polly Young-Eisendrath, and Terence Dawson, eds., *The Cambridge Companion to Jung* (New York: Cambridge University Press, 1995).

66. MK2 SA, CED Productions, France 3 Cinema, Cab Productions, TOR Productions, France, 1993, 1994, and 1994 respectively.

67. Sheldrake, *Seven Experiments that Could Change the World*.

68. Meier, *Wolfgang Pauli y Carl G. Jung*, 145.

69. Henrik Ibsen, "An Enemy of the People," in *An Enemy of the People, The Wild Duck, Rosmersholm* (New York: Oxford University Press, 1999).

70. Carl G. Jung, *Sobre el Fenómeno del Espiritu en el Arte y en la Ciencia* (Madrid: Trotta, 1999), 75. English version: Carl G. Jung, *Spirit in Man, Art, and Literature* (London: Routledge and Kegan Paul, 1975).

71. Jung, *Sobre el Fenómeno*, 87.

72. Circlot includes this quote by Baudelaire in his "La Séraphita de Balzac" *Confidencias Literarias*, p. 73.

73. ("Los 'sentimientos imaginarios' resultan . . . no en un nivel inferior de los sentimientos normales, sino superior.") Juan Eduardo Cirlot, "Los Sentimientos Imaginarios," in *Confidencias Literarias*, 102.

74. ("Si uno de sus dominios [del hombre] es la superficie del planeta, el otro consiste en las profundidades de su vida.") Ibid., 57.

75. Juan Eduardo Cirlot, "Los Elementos Imaginarios," in *88 Sueños, Los Sentimientos Imaginarios y otros Artículos* (Madrid: Moreno-Ávila Editores, 1988).

76. Juan Eduardo Cirlot, "El pensamiento de Gérad de Nerval," *Confidencias Literarias*, 98–99.

77. Juan Eduardo Cirlot, "La Mirada Humana," in *Confidencias Literarias*, 118.

78. Juan Eduardo Cirlot, *Diccionario de Símbolos* (Barcelona: Labor, 1994).

79. Cirlot, "La Mirada Humana," 66.

80. Lawrence Axelrod, *Cassandra Speaks* (tone poem for orchestra and electronic tape), Slovak Radio Symphony Orchestra, Szymon Kawalla (director), Vienna Modern Masters Label, 1999 (CD # 3012.) For more information, see Axelrod's Web page at ‹http://hometown.aol.com/larry223/index.html›.

81. Leonard Shlain, *Catalogue for "Antonio Arellanes"* (Laguna Beach, CA: Time Space Contemporary Art Gallery). The catalog is not dated but the exhibition took place in 1999.

82. See José Ortega y Gasset, *The Dehumanization of Art, and Other Essays on Art, Culture, and Literature* (Princeton, NJ: Princeton University Press, 1968).

83. "Death is a corpse. The external form we see is, in its interior, light. Death is a light that eternally says "no" while it actually means "yes." The beauty of the earth and the sky finds refuge in death" ("La muerte es una muerte. Vemos exteriormente cómo forma lo que, en el interior, es luz. La muerte es aquella luz que dice no eternamente, siendo si. En ella se refugia la belleza de la tierra y la belleza del cielo.") Juan Eduardo Cirlot, "Bronwyn IV." In *Obra Poética*, 247.

84. "Through Baudelaire and Rilke, we have gotten to know that beauty is all that is terrible and cannot be endured" ("Sabemos ya con Baudelaire y Rilke que la belleza es lo terrible que no se puede soportar.") Juan Eduardo Cirlot, "Unidad de la Poesía a Través del Tiempo," in *Confidencias Literarias*, 92.

85. Following Julia Kristeva's directives, we can say the language is studied by means of a "genotext" that is in charge of unraveling the ideological and unconscious mechanisms through which the text itself is generated. See Julia Kristeva, *Revolution in Poetic Language* (New York: Columbia University Press, 1984).

86. Ibid.

87. ("Me deslicé dentro de la cadena de salones comunicantes de la Gran Manzana. El gentío aparecía nimbado por el prodigio de las luces ocultas en falsos techos. La música y los murmullos se enroscaban en las profundidades cerebrales. Todas la criaturas se buscaban ansiosas de inyectarse el lenguaje tópico.") Miguel Ángel Diéguez, *En la Gran Manzana* (Alicante: Epígono, 1997), 1.

88. Davies, *Superfuerza*, 23.

89. See Gregorio Morales, "La Diferencia Cuántica," in *El Cadáver de Balzac*, 66–83.

90. Morales, *El Cadáver de Balzac*.

91. Quantum Aesthetics Group, *Quantum Aesthetics Group's E-m@ilfesto*, 20 October 2000 ‹http://teleline.terra.es/personal/lucschok/estetica/emailfestoeng.htm›.

92. I could provide exact references for these accusations, but I prefer that they be forgotten. I understand this is not much of a scientific attitude, but I am a writer, and a quantum one at that, and thus I do not want to take advantage of these smears. As the old Spanish saying puts it, "we can say the sin but never the sinner."

93. I will also not reveal how this and other anonymous messages got to me.

94. Jennifer Wilson, "Una Neva Estética ante el Próximo Milenio," *Turia* 46 (November 1998):304–306.

95. He is referring to the program created with the intention of giving apes the same rights human enjoy. For example, see ‹http://www.greatapeproject.org/ gaphome.html›.

96. Wilson, "Una Nueva Estética ante el Próximo Milenio," 305.

97. José Luis Serrano, *Validez y Vigencia: La Aportación Garantista a la Teoría de la Norma Jurídica* (Madrid: Trotta, 1999), 101–108.

98. Ibid., emphasis added.

CHAPTER 2

Quantum Language[1]

Mihaela Dvorac

INTRODUCTION

The radical paradigm change that is brought about by quantum aesthetics threatens profoundly the general conception of language, transforming speech substantially and differentiating it from the movements that characterize the twentieth century, specifically formalism and structuralism. In any case, as we have seen, in some respects the view of language linked to quantum aesthetics is similar to that associated with the transformational grammar of Noam Chomsky, particularly the distinction he makes between deep and surface structure, although this relationship is minimal. The quantum paradigm revolutionized language at a time of social crisis, particularly related to the nature of speech and the subject who speaks.

Although the study of language has advanced significantly by overcoming the shortcomings of the Newtonian paradigm, we can say that quantum aesthetics goes beyond the views of language operative during this earlier period by articulating a perspective whereby each linguistic element is explained and integrated into the entire speech system. This was not the case, for example, in the most important theories of the twentieth century, which were basically offshoots of the theories produced by positivists and Saussurians. For example, while returning to Chomsky, language has more to do with a fixed system of rules, through which speakers transform their thoughts into statements, rather than the study of the unconscious side of speech.

This eradication of linguistic meaning was expressed most clearly by Louis Hjelmslev, who ignored semantics by reducing language to a mechanism that merely records events.[2] This is the thrust of his theory of language that is known as "Glossematics." Consistent with other conclusions of quantum aesthetics, this approach to language must not be seen as simply naive, because his thesis overlooks the complex and difficult facets of language. Hjelmslev's primary interest was to invent a science of language that is similar to algebra or normative logic.[3] His focus, therefore, was the discovery of deductive linguistic principles that, in the end, are divorced from everyday activities and experiences. What Hjelmslev proposed eventually is a purely formal and lifeless model to describe language.

Quantum aesthetics is clearly connected with other attempts to study language that are more comprehensive, such as the program provided during the 1970s by the group Tel Quel.[4] According to these critics, the crisis or "rupture" in language probably occurred at the end of the nineteenth century and brought about three significant changes. First, the members of Tel Quel began to address the possibility that language has various dimensions. Second, they tied language use to several key social factors; language, in other words, was no longer simply a self-contained and abstract system. Linguistic expressions, as proponents of Tel Quel liked to say, existed at a nexus of historical forces. And third, language use and psychoanalysis were linked together, in order to emphasize that speech is mediated by subjectivity. A quantum conception of language tries to develop and integrate these three levels of language use.

THE IMPORTANCE OF THE HUMAN

In order to begin a proper assessment of linguistics, one must not make the customary generalizations about language, as if the differences that exist are merely imperfections that, in effect, testify to the existence of universal norms. With respect to quantum aesthetics, the theses of "fuzzy language" and "individuation" have made evident the importance of variations in language use and how these differences nurture the development of free and creative persons who attempt to know what separates them from their cohort, nation, or family. These individuals search for their so-called specificity.[5] Supporters of quantum aesthetics believe that any attempt to eradicate the speaking subject is not only condemned to fail, but also fosters a serious misperception about the nature of language. The focus of the quantum approach to studying language, therefore, is the subject who speaks.

Recently, scientists have begun to understand that they cannot avoid affecting the elements that they study. Their experiments, in other words, are not neutral but shape their perceptions of events. For example, this finding was an important outgrowth of the so-called Hawthorne experiments, whereby the presence of researchers affected the responses of the persons who were under investigation. This phenomenon came to be known at the "Hawthorn effect."[6] In the field of physics, a similar event occurred. When researchers tried to identify the spin or rotation of neutrons with respect to a fixed reference, the rate of movement simply corresponded to the movement of the referent.[7] Accordingly, identical to the way scientists project themselves into their experiments, speakers introduce themselves into their use of language. The result of this intrusion is that language provides the subjective view that persons or groups have of reality.

Every text is written in a language, but the linguistic text might be equivalent to the "visible material" that surrounds persons and that each individual is able to mold according to his or her psyche, interests, or intellectual heritage. That is, the speaker arranges this material, supplies it with form, and resides in the midst of the process of speaking. Therefore, language, or text, is not a representation of something that exists prior to speech, but rather simultaneously expresses itself and the various, and often parallel, modes of reality that comprise the world.

THE TEXT AS A POINT OF DEPARTURE

The revolution in understanding language that has occurred in the past few decades has historical, social, and theoretical implications that have been manifested in the notion of the text. For example, literature begins to be principally language, or is perceived to be language. The influence of structuralism in the human sciences has moved frequently toward a convergence of stylistics, poetics, and literary semiotics. But the interest of linguists has focused increasingly on how a text functions and, above all, is produced.

Those who advance a quantum conception of language consider literature to be a text. Nonetheless, from the time of Ferdinand de Saussure until today, literature has been considered to be a system of signs. Within the framework of semiotics, literature has been treated as an important practice similar to scientific studies or journalism, but, nonetheless, different from them because literature allows for the actual production of a text to be understood.

According to Saussure, language is a system of signs that expresses ideas; the significative power of a sign resides in the relation that is established between the signifier and the signified, which are the two inseparable sides of a sign. But this Saussurian view of the sign cannot explain how a text is produced. Despite the apparent symmetry that exists between the signifier and what is signified, subsequent to defining language to be a system of signs that represents ideas, any sense of equity between these two elements is lost. That is, the assumption is made that something autonomous exists that is represented and expressed by the sign. Saussure's rendition of the sign, therefore, provides little insight into the actual production of a text. In many respects, his signs are inert things.

The new reflections on the nature of the sign by members of the group Tel Quel, and also by the Prague Circle, have begun to reveal that poetic language departs appreciably from the traditional models of the sign. Specifically, literary signs are not identical to words and should not be differentiated into two parts, the signifier and the signified. Such a descriptive system embalms language. By substituting this notion of the sign for a minimal conjunction of symbols, including the logic of the sign and the text, language can be understood to be more than a depository of meaning. Now, language can be envisioned to produce meaning.

The significant change in perspective from viewing language as merely a functional facet of a text reveals the problem of separating poetic from regular language. Specifically, poetic language is presumed to represent a deviation from the usual and responsible way of representing the accepted reality. In this regard, poetry appears to be a method of exploring the possible ways of expressing an idea or event, with each attempt becoming increasingly profound. Given this experience, regular language seems to become more and more restricted and, possibly, something to be avoided.

What confers poeticality on language is precisely the ability of persons simultaneously to make themselves and give the world meaning through the exercise of speech. Poetry, in this sense, is consistent with the theory of language held by Tel Quel. These writers argued that history is produced, and not reflected, in a text.[8] Writing, in other words, is a process of production and not representation. For the critics associated with Tel Quel, poetry is not tied to reality, but rather is an activity of experimentation; poetry sets free the linguistic code. As a result, words are used in novel ways that engender diverse experiences and insights, otherwise known as multiple realities.

LANGUAGE AND THE UNCONSCIOUS

In quantum literature, the structure of language is recognized to be located in the unconscious, where the self is able to define itself as a signifier. The body of theoretical work developed by Sigmund Freud and Carl G. Jung has expanded the notion of reality to include even the unconscious. Thus, a text does not merely refer to an autonomous and objective reality, but instead refers to a reality that persons compose or construct through their own perceptual activity. With reality extended to the unconscious, the text becomes an expression of this domain. In other words, according to quantum theory, a text cannot refer to an objective reality that is outside of persons, but instead is invented by them. The text, according to Roland Barthes, is an expression of ambitions or desires that become real through language. The meaning of language operates under or behind the empirical components of speech. Rather than clarifying codes, the purpose of studying speech is to reveal expressions that are not always clear or certain.[9]

At the same time that an infinite number of universes converge in persons, they have the ability to select one or more, and also have the ability to create one of their own. In this sense, perception is not only selective but constitutive. That is, the same "anthropic principle" that operates in physics also is relevant to language. This principle, writes Gregorio Morales, is based on the idea that human perception is intertwined with the universe in such a way that the resulting product should be known as mental matter.[10] This is another way of describing the Hawthorne effect. This effect taps into the idea that scientific and mathematical formulations resemble their creators. Humans, in other words, are the mirror in which the universe is reflected! One way or the other, what seems indisputable is that human perception is connected to the cosmos. So much so, that the products of both are interchangeable. This is why Arthur Eddington can go as far as saying that the universe is constituted of mindful matter.[11] Reality, stated simply, is connected intimately to the rhythm of personal and collective existence. The existence of different idioms and their corresponding realities, for example, demonstrates that each one is valid and true at the same time. None implodes automatically because of a lack of veracity.

WHAT IS INSIDE OF THE TEXT?

When known as a body of signifying practices, a text is defined as a product that is expressed inside of and through a system of norms that is linguistic. The fundamental dimension of a text is productivity, and thus through a process of structuration the meaning of this narrative and the speaking subject are articulated. In other words, the text represents the exercise of language.

In order to enter a text, the distinction made by Julia Kristeva between the phenotext and genotext might be helpful.[12] The genotext underlies the text—and conveys its expressive or intended meaning—and is more real than the surface statements that are made. The genotext is constituted by the rules through which expression is transformed into the statements that comprise a text. As will be seen, similarities with the transformational grammar of Chomsky are evident, although Kristeva does take a much stronger stand on issues such as meaning, semantics, content, and discourse. From the perspective of quantum aesthetics, Kristeva has proposed an integrated and complex rendition of language. Indeed, her distinction between phenotype and genotype is significant in illustrating a quantum view of a text.

Kristeva's key concern is that language be understood as more than strings of words that are united by rules. A text, in other words, should no longer simply be envisioned as a mechanism used to convey meaning. Much more important, she claims, is that every text is underpinned by a process that engenders meaning or the genotext.[13] If the surface rules are the focus of attention, a reader will miss the process whereby a text is given life. Moreover, the speaking subject will be obscured behind a wall of meaningless objects. In a quantum sense, the genotext represents authors projecting themselves into and organizing their texts.

From this point of view, the phenotext is something that the reader or critic must go beyond in order to arrive at the deeper meaning of a text or the genotext. The phenotext-genotext link is the same that exists between the so-called signified structure and significant productivity. Whereas the signified structure has an objective appearance for the speaker, the activity of producing a text is not. Accordingly, the specificity of a text resides in translating the thoughts—which are modeled after language and thus are inseparable from it—into something observable, or, in other words, in the process whereby the rules of the genotext produce a readable text. A text is thus something fragile, because the meaning that emerges from the genotext is dynamic and can be transformed.

The difference between realism and quantum aesthetics is that realists understand texts to be objective. Quantum writers, however, look beyond this level to discover the significance of a text. The subjective view of reality, which also includes persons, is built from certain hidden laws that explain how a text acquires meaning replete with various influences and particular "ideograms."[14] For quantum writers, literature is not simply a reflection, or nothing more than an image, but instead represents a speaking subject and a creative linguistic community. In addition to recognizing the two traditional sides of a novel—as linguistic narration and discursive fact—Kristeva identifies a third aspect with which quantum aesthetics is in agreement. That is, a text is produced and continually reworked.

A quantum critic enters a text in order to reveal it or, stated differently, to make known the genotext; these writers want to unmask the apparent reality of a text, along with its ideological features. In other words, what a quantum writer does is to search for these ideograms and advance beyond the lexical, semantic, and syntactic facets of a text. The purpose of this change is to reestablish its existential direction and significance. Someone with a quantum orientation tries to perceive and describe the parallel realities that converge in the genotext. Consequently, a text is presumed to have a double orientation: a text embodies both a particular linguistic praxis and social process. Furthermore, every text is criss-crossed by other texts, which affects how any one is read. Kristeva refers to this phenomenon as "intertextuality."[15] This association permits, for example, any one text to be interpreted in view of others, and thus can be given a more profound significance. This practice has no final denouement—the transposition of any one text can assume many forms and is never complete.

THE VALUE OF THE SYMBOL

While the realists-positivists are interested in the sign, quantum theorists are interested in the symbol. In this case, symbols are understood to filter reality and underpin the surface structure of a text. As might be expected, quantum art will include the transformation of this surface or apparent reality.

Quantum artists break with the ideology of representation and thus no longer deal with signs. After all, signs merely mimic reality. Instead, these writers and painters utilize symbols, as opposed to signs, that condition and encourage reality to emerge in various forms. A symbol, accordingly, unites many contradictory elements; a symbol contains and releases various parallel realities. A symbol can be considered to be a cosmogonic semiotic practice: that is, symbols promote the transcendence of the surface of texts that does not terminate in any particular foundation. Symbols allow the meaning of a text to proliferate.

Symbols possess multiple meanings that are imprecise and full of possibilities. For example, mythic thought imbues a symbol with many dimensions—

heroism, fear, love, virtue, and solitude—that are simultaneously present. The function of the symbol, therefore, is to specify which of these dimensions is more or less important at any one time. As opposed to the sign, the symbol restricts meaning in this manner, while also creating meaning.

The rendition of the symbol that a quantum writer supports operates on the basis of the continuity present between the signifier and what is signified. Similar to the sign, the symbol is expressive; as opposed to the sign, however, a symbol does not refer to a single, autonomous reality, but rather evokes a collection of images and ideas. Saussure defines the symbol in the following manner: "One characteristic of a symbol is that it is never wholly arbitrary, for there is a rudiment of a natural bond between the signifier and the signified."[16]

Of course, this relationship is recognized by supporters of quantum aesthetics, but they add another very important feature to the symbol. They argue that the symbol represents a reciprocal relationship between the signifier and signified, a synthesis of opposites. Their point, as Morales suggests, is that symbols are not restricted by reality and thus join, simultaneously, a host of possible realities.[17] The signifier is certainly related to reality, but a reality that is generated through symbolism that can be reworked in almost unlimited ways. Quantum artists claim that subjectivity should not make a fetish of objectivity, or the power and importance of the symbol will be lost.

Proponents of quantum aesthetics include in their work the various dimensions that they believe symbolism opens to the unconscious. This modus operandi is consistent with their elevation of the genotext in importance. When investigating this supposedly unknown, hidden, and impenetrable aspect of reality—a facet often viewed as more real than what is "believed to be real"—a quantum writer uses symbols as usual, but in a manner that is Jungian, not Freudian, and very complex. Quantum symbolism is completely new, although the roots of this approach are found in the literary movements at the end of the nineteenth century and the beginning of the twentieth. Although the symbolism used by quantum artists may appear similar to that utilized throughout literary history, there is a new characteristic: these symbols are thought to tap into the individual and collective unconscious. Through the use of symbols, these artists mold a deep reality by separating the irreal from the real, or the surface structures from the existential meaning of a text.

For a poet such as Rosario de Gorostegui, the act of creation—poetry—signifies "the transformation of a particular vision into another that is kaleidoscopic and complex." The aim of poetry should be, then, "to convert the hidden interior (of a person) into universal metaphors."[18] In other words, the task of a writer is to make the particularity of the genotext meaningful for a variety of persons. Regardless, Gorostegui's intimate poetry is a real song not only to love and the joy of life, but also to the fullness of the moment. For example, she writes: "Splash me with aromatic herbs / dream of a non-repeatable history of love. . . . I prepare an infusion of dreams / so that you read the future in the sediments of my absence.... Give me red poppies and luck /in order to invent the future that you foretell."[19]

The word "dream" is reiterated in this poem in order to reveal a reality that is formed in the unconscious. These dreams, such as a "racism of fleeting dreams," the "living the dreams of the magic wand," or the "infusion of dreams," are not anchored in the apparent reality, but in a deeper and more truthful one. In these examples, the usual understanding of dream is redeployed to suggest a critique of reality that comes from an alternate source.

For Miguel Ángel Contreras, poetry is "a form of perceiving the world; a mode of interpreting life from the perspective of allegory, myth, and emotion."[20] The origin of his creative experience is recollection and experience. Also, he recognizes that his descent to the depths of the psyche is the act of creation. In this regard, the symbol in the poetry of Contreras is close to allegory. For example, the desert covered with sand dunes represents the hollowness of human beings in the modern world. As he says, "desert, everywhere is desert / and my body / a physical extension / of my eternal internal desert.... Desert and solitude / they accompany and prolong each other."[21] Other times, the images of the dusty and sterile desert is a metaphor for the illusion of love: "How can I find you/in this immense patch of sand / that is the desert that covers you.... How can I find your face."[22]

The poetry of Francisco Plata constitutes a fascinating and focused body of work. For him, poetry is a "multifaceted contribution,"[23] and his verses are an outgrowth of his meditation on the mystery of life. "Smelling like death you come down / You go down the street with the smell of human decay / And your arms, shoulders, and sagging skin drop off of you.... Three streetlights are thrown at you that invite you. / Death on your heels already.... / And you continue living."[24] Death, the central mystery of existence, is contemplated by Plata from the position of an observer, who is involved in uncovering the enigmas of life. He is not merely recording or documenting these mysteries, but is reenacting them. For example, more than simply a fact of life, death is a process that all persons must navigate in their own way. They must read or interpret this event alone and uniquely.

According to Morales, a novel should reveal the unconscious dimensions of its characters, at least its key characters. In this way, he considers the unconscious to be a text that can and should be read; the novel, as a literary work, expresses the psyche of the characters and reveals the norms that guide their thoughts and actions, that is, the genotext that lies inside of people. This is the theme of his novel *La Individuación*.[25] The protagonist of this work initiates a trip not to discover the external reality of the world, but rather to expose the masters who rule his consciousness. The purpose of this voyage is to gain insight into the source of his thoughts, actions, and, taken together, his biography. This is more of a trip through time than space. And, as Morales suggests, the solutions to the enigmas of daily life reside within the person, rather than in society or other individuals. *La Individuación* is thus a peculiar detective story that emphasizes self-interrogation as key to resolving problems.

The novel by Miguel Ángel Diéguez, *En la Gran Manzana*,[26] is also an investigation into the hidden levels of a text. As part of this trek, the author criticizes the alleged objective reality in this story and reveals this domain to consist of various worldviews. Diéguez unmasks the language and mode of narration and, in a manner similar to Morales, becomes associated with the generation of the text. In this work, the readers are part of a process whereby the ideological and mental categories hidden in the text are revealed. Diéguez writes: "I slid into the chain of contiguous rooms in the Big Apple. The crowd appeared to be surrounded by a halo created by the extraordinary lights in the false roof. . . . All of the creatures searched themselves, anxious to get a fix of topical language."[27]

The author enters this topical—superficial—language and begins to interrogate it through the actions of his main character Ricardo Verela. In this book, the protagonist struggles against the "scavengers" who have infiltrated the computer network of the Big Apple and remakes language with the aim of "decomposing reality." The entire struggle in this novel is over language, between those who know the enormous power of a message and those who ingeniously attach this power to reality. The author's central idea is that any attempt to remake language also changes the reality of those who use such speech. Those who control language are able to shape reality.

ARCHETYPES AND MORPHOGENIC FIELDS

A multitude of symbols form an archetype or a morphogenic field.[28] Without a doubt, the reader has become familiar with these terms because of their appearance in other chapters of this volume. Regardless, a creator of quantum literature is extremely interested in the archetype, according to which the heroes of this genre live. This is what occurs to Gabriel Alvis, an important character in Morales's book *La Individuación*. Alviz lives unconsciously according to an archetype and, in this sense, is dominated by this imagery. Only at the moment when he is able to grasp and understand this archetype does he find true freedom or liberation. Said another way, Alviz needs to "to read himself as an autonomous person."

A study of the so-called folded order that is hidden and indecipherable—an order like the genotext that gives daily existence meaning—leads the quantum author to the archetype that lives in an individual. Coming to know this image, persons can release themselves from it and begin their journey to freedom. For quantum theorists, an archetype is problematic only when a person is unaware of this phenomenon. Through self-exploration, persons become masters of their various life-options that converge in an archetype. In the manner intended by Socrates, quantum artists believe than an unexamined life is not very meaningful. Or in a manner similar to Immanuel Kant, a person who lacks self-reflection can never be free. As a consequence of this sort of reflection, individuals can become individuated; similar to a text, they can write themselves in many different ways.

THE QUANTUM WRITER

A quantum writer is not forced to say something about the external world. Due to the dogma that a writer must say something, the old realist and positivist art conceals the fact that texts are produced. The focus of attention, accordingly, is not the process of production but the product of artistic activity. And consistent with the pragmatic nature of the traditional dogma, artworks are merely consumed without any interest in the creativity or imagination that generated them.

Quantum aesthetics understands writers to have two basic traits. First, they are implicated in language. And second, they are theorists who explore language, the world, and themselves. To be more specific, these writers break with what has come to be known as the "representational thesis," and thus their work is not mimetic.[29] Instead, they open the text in a way that leads to true reading. Their approach to reading might even be called radical, in that readers are encouraged to invent a text as they read. Texts are never simply encountered, or reread, but are read anew each time they are entered.

CONCLUSION

As can be seen, a quantum conception of language has a primordial objective of acquiring self-knowledge and advances the project of individuation. If persons study language, the aim of this endeavor is to know themselves as they define themselves and the world. Accordingly, language is not conceived to be a science that is severed from the speaking subject, but rather is something that must be penetrated in order to understand human complexity.

Moreover, quantum theorists do not reject the idea that at the root of all language is a spirit or soul; all linguistic expressions, in other words, embody unique and nonrepeatable vibrations that constitute the most essential element of every human being. This is especially true for those persons who are advancing toward individuation. Quantum aesthetics, then, is centered fundamentally on the language of the individual and the differences between this mode of speech and the language of the family, group, or community. From this perspective, language is an element that is distinguished from narrative, and thus quantum writers search for the rhythm or the musical interior of language.

NOTES

1. Translated into English by John W. Murphy.
2. Louis Hjelmslev, *Prolegomena to a Theory of Language* (Bloomington: Indiana University Press, 1953).
3. Robert de Beaugrande, *Linguistic Theory* (London: Longman, 1991), 124.

4. Mary Ann Caws, ed., *About French Poetry from DADA to "Tel Quel"* (Detroit: Wayne State University Press, 1974), 90. Also see Patrick Ffrench, *The Time of Theory: A History of Tel Quel (1960–1983)* (Oxford: Clarendon, 1995).

5. Gregorio Morales, *El Cadáver de Balzac* (Alicante: Epígono, 1998), 9.

6. See Rupert Sheldrake, *Seven Experiments That Could Change the World* (London: Fourth Estate Limited, 1994).

7. Paul Davies, *Superfuerza* (Barcelona: Salvat, 1994), 30.

8. Toril Moi, Introduction to *The Kristeva Reader* (New York: Columbia University Press, 1986), 1–22.

9. Roland Barthes, "From *The Pleasure of the Text*." In *A Barthes Reader*, ed. Susan Sontag (New York: Hill and Wang, 1982), 413.

10. Morales, *El Cadáver de Balzac*, 22.

11. Ibid.

12. Julia Kristeva, *Revolution in Poetic Language* (New York: Columbia University Press, 1984), 86–87.

13. John Lecht, *Julia Kristeva* (London: Routledge, 1990), 128.

14. Julia Kristeva, *Language: The Unknown* (New York: Columbia University Press, 1989), 28.

15. Kristeva, *Revolution in Poetic Language*, 59–60.

16. Ferdinand de Saussure, *Course in General Linguistics* (New York: Philosophical Library, 1959), 68.

17. Morales, *El Cadáver de Balzac*, 100.

18. Fernando de Villena, ed., *La Poesía que Llega: Jóvenes Poetas Españoles* (Madrid: Huera and Fierro, 1998), 31.

19. Gorostegui, Rosario de, "El Futuro que Adivinas." in *La Poesía que Llega: Jóvenes Poetas Españoles*, eds. Fernando de Villena (Madrid: Huerga y Fierro, 1998), 33.

20. de Villena, *La Poesía que Llega*, 79.

21. Contreras, Miguel Ángel, "En el desierto," in *La Poesía que Llega: Jóvenes Poetas Españoles*, ed. Fernando de Villena (Madrid: Huerga y Fierro, 1998), 83.

22. Ibid.

23. de Villena, *La Poesía que Llega*, 107.

24. Plata, Francisco, "Olor a muerte," in *La Poesía que Llega: Jóvenes Poetas Españoles*, ed. Fernando de Villena (Madrid: Huerga y Fierro, 1998), 109.

25. Morales, *El Cadáver de Balzac*, 36–52.

26. Miguel Ángel Diéguez, *En la Gran Manzana* (Alicante: Epígono, 1998).

27. ("Me deslicé dentro de la cadena de salones comunicantes de la Gran Manzana. El gentío aparecía nimbado por el prodigio de las luces ocultas en falsos techos . . . Todas las criaturas se buscaban ansiosas de inyectarse el lenguaje tópico.") Ibid., 1.

28. Morales, *El Cadáver de Balzac*, 38, 107.

29. Richard Rorty, *Objectivism, Relativism, and Truth* (Cambridge: Cambridge University Press, 1991), 4.

CHAPTER 3

Quantum Literature[1]

Francisco Javier Peñas-Bermejo

INTRODUCTION

During the seventeenth and eighteenth centuries, advancements in the study of the observable aspects of empirical reality became evident. With the development of calculus and analytic geometry, a link between the elements of theoretical physics and the physical world was established. With this turn of events, quantitative methodology became an essential tool for the elaboration of subsequent hypotheses. Rationalism, the first branch of modern philosophy, initially advanced by René Descartes (1596–1650), asserted on the one hand the autonomy of reason—which becomes the only and supreme basis for knowledge—and on the other hand, the use of mathematics (specifically, analytic geometry) as the language that best connects experience and nature. Furthermore, Descartes established a dualistic structure of reality that influenced decisively subsequent scientific reasoning. The mechanistic and deterministic notion of the material world proposed by classical physics was advanced through Isaac Newton's (1642–1727) invention of *infinitesimal* calculus, the theory of gravitation, and the laws of mechanics, in addition to the application of magnetic and electrical forces to the study of reality. These developments constituted the pillars of modern science.[2] The success of positivism in the nineteenth century, with Auguste Comte (1798–1857) as its main proponent, furthered the decisive role of science and reified the tenets of past centuries concerning the rigorous and deterministic methodology used in assessing reality.

Understanding the physical universe took a new turn during the beginning of the twentieth century, after Albert Einstein (1879–1955) revised the mechanistic model of Newtonian science that dominated the eighteenth and nineteenth centuries. Einstein's theoretical proposals about (1) the general and special relativity regarding a tetra-dimensional space-time continuum that is considered finite but unlimited and curved, (2) his equation regarding the conversion of mass into energy when the quadrate of the speed of light was reached, (3) his quantum proposals regarding the effect of photoelectricity, (4) his effort in systematizing a unified camp of electromagnetic and nuclear forces, and (5) the subsequent philosophical implications that derived from his work modified the conventional modes of thinking regarding physical phenomena.[3] Within thermodynamics, Max Planck (1858–1947) investigated the relationship between heat and electromagnetic radiation, thereby overturning the predictions of classical physics by demonstrating that atomic-level energy (light, radiation, heat) is emitted in energy packages called quanta that integrate wavelength, frequency, and energy harmoniously through "Planck's constant" (h). For the first time, there was irrefutable proof of the discontinuity of reality. Subsequently, Louis-Victor de Broglie (1892–1987) demonstrated that light is manifested both as a wave and a particle. The atomic model proposed by Niels Bohr (1885–1962) suggested the need to abandon classical ideals about formulating complete causal explanations of physical phenomena, and accordingly revealed the appropriateness of statistical explanations. According to Bohr, the discontinuous behavior of material reality is not evident until sufficiently small systems are examined. From this position, Bohr began to develop the notion of *complementarity*. Despite their exclusionary and contradictory ideas, only by integrating in one explanation (1) the duality of light, (2) the determinism of classical physics in macroscopic systems, and (3) the indetermination of microscopic systems found in quantum mechanics can we have a good description of the nature of physical reality. For Erwin Schrödinger (1887–1961), matter was analogous to light, because they are both a wave and a particle. He elaborated the fundamental basis of mathematical wave mechanics wherein matter is not sometimes a wave and other times a particle, but rather simultaneously a wave and a particle.

However, Werner Heisenberg (1901–1976) may be the main contributor to the quantum revolution of physics in the twentieth century due to his development of the principle of uncertainty or indetermination. Contrary to the Newtonian view of the world, Heisenberg demonstrated that precise mathematical descriptions of reality would always be impossible, not because of technical or methodological problems that may be overcome, but due to existential reasons that are impossible to surmount. Based on mechanical matrices similar to Schrödinger's wave equations, Heisenberg proposed the first widely applicable quantum theory. This theory was based on the fact that the interaction between observer and object causes uncontrollable alterations in the observed system as a result of the discontinuous changes characteristic of subatomic processes. Likewise, he demonstrated that it is impossible to determine the position and the momentum of an electron simultaneously. Later,

Heisenberg accepted the value of Schrödinger's wave mechanics and Bohr's principle of complementarity, and supported the Copenhagen version of quantum mechanics. According to this school of thought, the classical notions of causality and determinism must be abandoned in favor of indetermination, complementarity, and subjectivity.[4]

SOME QUANTUM INFLUENCES

The ultimate implications of the quantum paradigm have yet to be established in a definitive format. Nonetheless, quantum interpretations have indeed affected the way in which we relate to the world, as well as the way we understand the connections among the various disciplines. In *The Conscious Universe: Part and Whole in Modern Physical Theory*, Menas Kafatos and Robert Nadeau emphasize Bohr's principle of complementarity. They indicate that this principle underpins the universe and is not only present in physical theories of the world, but also in the rendition of human existence that attempts to overcome dualism.[5]

Quantum influence has extended to other disciplines. In the field of education, for example, Bruce Wilshire recognizes that the university is an undertaking that integrates the self and the world.[6] In management science and organization theory, Dick Richards[7] and Margaret Wheatley[8] have made quantum implications explicit. In medicine, Deepak Chopra discusses *quantum healing*.[9] In the area of psychology, the famous friendship between the psychologist Carl G. Jung (1875–1961) and the quantum physicist Wolfgang Pauli (1900–1958), as well as the coincidence between their respective theories, is well known.[10] And in the field of politics, Flora Lewis asserts that uncertainty is "a value that requires respect for and a degree of deference to the observations and ideas of other people."[11] In this sense, connections can be established with multiculturalism.[12] For multiculturalists, the diverse cultures and ideas present in all the spheres of life have provided fruitful opportunities for research in many disciplines, especially postmodern literary criticism. The urge to dissolve fragmentation and dualism in favor of an ever-expanding contextualism is a solid proposal that José Luis Gómez-Martínez presents in his book *Más Allá de la Pos-Modernidad: El Discurso Antrópico y su Praxis en la Cultura Ibeoamericana (Beyond Postmodenity: The Anthropic Discourse and Its Praxis in Latin American Culture.)*[13]

EXAMPLES OF QUANTUM LITERATURE

As the effects of subatomic physics are assimilated into an era dominated by technology, information, communicative speed, and the tendency toward globalization, both science and art begin to assume and incorporate the new paradigm into their interpretation of the world. In this sense, we can talk about a type of physics that has acquired a commitment and a responsibility for producing social and political change similar to Jean-Paul Sartre's "engaged literature."[14] Quantum notions have been the objects of personal reflection and renewal for writers for quite some time.[15] The modern "cosmo-vision" advanced by quantum physics has contributed to fostering an alternative conception of the narrative and artistic universe, as Manuel Garcia Viñó has demonstrated in "La Novela Relativista y Cuántica: Materiales para la Construcción de una Teoría Aplicable a Otras Artes" ("Relativist and Quantum Novel: Materials for the Construction of a Theory Applicable to Other Arts")[16] and as Gregorio Morales shows in *El Cadáver de Balzac (Balzac's Corpse)*.[17] Additionally, the Quantum Aesthetics Group was created in February 1999 and has published a manifesto on the Internet.[18]

The application of quantum theory to literature has attempted to create a space for freedom, extraordinariness, and imagination, where academic realism—that does not recognize the complexity of reality—is replaced by a version that is more transcendent. In the preamble to the "Declaración del Salón de Independientes,"[19] the renovation in literature proposed by quantum aesthetics is revealed: "We, as writers, think that for the purpose of mere survival we need to rebel against anything that has to do with the culture of power, the politics of control, and the false progress that tends toward a type of wealth that exploits our world and our minds and does not care about happiness."[20] Within the confines of this vindication of the individual and diversity, various authors have initiated a creative process that goes beyond immediate concerns in order to delve into human destiny. The following pages present a necessarily brief synthesis of quantum literature related to the Spanish novel and poetry. In order to do this, the various quantum principles mentioned earlier are used.

Quantum Novel

The rich intrigue in Manuel García Viñó's *El Puente de los Siglos (The Bridge of the Centuries)*[21] is grounded on the practical application of relativist and quantum theories that underpin the restless meanings that harass Arturo Hispano 9,502. A citation by Arthur Schopenhauer, "All types of art aspire to be music" ("Todas las artes apiran a la condición de la música"), prepares the reader to experience the novel as a whole that is comprises the continuous harmony of diverse parts, situations, conditions, and sentiments. All of these factors provoke in the reader a sense of strangeness that, at the end, is overcome when the secret parallelism between cosmic phenomena and human life is revealed.

The action of the novel, which takes place between the second and third millennium, is based on both a love triangle and a trial for a well-documented murder that actually did not occur. This action responds to different temporal states that simultaneously coincide on various occasions. *El Puente de los Siglos* introduces three theories. First, the behavior of an individual not only influences the future but also alters the past. After a two-year-long relationship, Rosy tells Arturo the following: "Everything we have built together you have destroyed in this very moment with your betrayal, lies, and cowardice. Our actions, Arturo, not only affect and determine our future; our current actions also modify our past."[22] The second theory introduced by García Viñó is the possibility of experiencing a temporary change in consciousness induced by a drug. Arturo is in the middle of a triple conjunction of (1) a mechanic medium that acts as a receptacle (himself, or any other person), (2) a drug, and (3) a hypnotic state. The physician confers with Arturo: "We have discovered a drug ... a drug, my friend—not a machine—a drug capable of disintegrating a human being and putting him back together in a pre-determined time in the maya world.[23] Of course, this point in time will be part of the temporal and perishable, albeit different, world. Ah! Because of our limited knowledge, this world can only be in the past."[24] The third idea put forth in *El Puente de los Siglos* deals with the issue that each person is the center of his/her own universe and relates to other human beings intersubjectively. The Hermano Mayor de la Raza (Big Brother of the Race) explains to Arturo: "Each of you, mortals, is the center of a Universe, his Universe, one of the infinite universes out there. Anything that can be imagined is possible.... Many of the events and things of your universe can be used by other mortals who live *simultaneously* with you, at least to a certain—sometimes important, sometimes unimportant—extent. These mortals' universes interfere with, and, thus, are not completely incompatible with yours."[25]

In *El Puente de los Siglos*, several principles are advanced, such as the elastic identity of time, the existence of parallel universes, the possibility of different states of consciousness, the revelation of a transcendent whole, the correspondence of the real and the possible, the integration of the parts and the whole, and the interpretive interrelationships of reality throughout history. The fundamental tenets of this novel both question the traditional understanding of what is known as reality and explore this source of knowledge. The novel concludes in a glorious finish when Arturo's death fulfills a personal destiny that opens his way to transcendence: "The dead, the non-born, the resuscitated, the alive understood that he was in the center of the world and that something had started anew in some other place."[26]

Gregorio Morales's novels are characterized by his impeccable mastery of language and narrative voices, as well as his passion for the discovery and enlightenment of human experience. The twenty-eight chapters of *La Cuarta Locura* (*The Fourth Madness*)[27] symbolize the ascension of the Luna-Amaterasu-Iria Clarós through various phases until reaching her fullness. Clarós is a ballerina whose life is recreated by Fermín Alzu with a set of photographs of her that he found by chance. As Alzu removes the veils from Salomé-Iria, both the fascination of finding the essence of beauty[28]—for seizing the infallible—and the vertigo caused by the unknown and the Dionysian chaos that dominates the characters become apparent. This chaos encompasses the foundation of beauty and can be seen in the artistic expression of the dance as truth, the reconciliation of the unconscious and conscious, and the personal search for individuation.[29] *La Cuarta Locura* is guided by the motto, "mystery plus difference." Through this postulate, quantum aesthetics advocates that the human being recognize his/her own mystery and complexity. Iria Clarós pursues this motto through a "more underground dance, a dance of darkness from which the most obscure domains of the self can arise."[30] The imaginative freedom of the novel vindicates (1) the human being's independence from mechanistic determinism, (2) subjectivity as the foundation for creating one's own universe,[31] and (3) intuition's fundamental role "as the only way of getting to the unknown."[32] This extraordinary novel also discusses the activation of certain domains of consciousness by provoking alternate or holotropic states that permit movement, from one part to the hologram (to the whole), that is, to the essence of beauty.[33] This beauty is finally revealed as Alzu tells Iria Clarós: "What have you pursued though your work, through the vertigo produced by your work throughout your life? Arriving at the intuition of death. That's precisely what your nudes show! You have pursued the tremors of death, which is the same as striving to know the essence of life."[34]

In the exciting novel by Morales, *El Pecado del Adivino* (*A Fortune-Teller's Sin*),[35] Adrián's oracles announce a reality that is simultaneously "A" and "not-A," as advocates of complex thought and of quantum physics claim is possible based on the principles of uncertainty and complementarity.[36] The wide range of possibilities that unfold before each of Adrián's predictions, along with chance, coordinate the uncertainty involved in integrating the beginning with the end and/or the parts with the whole. This interpenetration is not only produced by the plot, but also by the broad narrative perspectives that assume the continuity of space and time. The process of individuation is the dominant message. Adrián, for example, abandons his law office for a psychic studio. This change allows him to learn about himself and realize that he is actually looking for Ángela.[37] The same goes for Mónica, who, in a surprising end, finds in Gaby an experience of an ancestral past. The combination of the possible and the unstable, the interchange of time and events, and synchronicities[38] produce a world in which the ephemeral and the permanent, past and future, passion and rationality, instinct and frigidity, the concrete and the uncertain, and the artistic wonder and the gloomy crypt are reconciled. These unions finally integrate the hidden roots of humans and allow them to "be the universe, not just an insignificant part of it."[39]

The action in the disturbing novel *Los Días del Duopolio* (*The Days of the Duopoly*)[40] by Miguel Ángel Diéquez deals with the growing limitations placed on human freedom and imagination in a society controlled by structures of power, technocracy, the manipulation of information, and standards of social behavior that are imposed on everyone. The conflict between the internal and external reality of the protagonist Julio Abril is unleashed in a spontaneous manner that drags him into a state of instability whereby he loses sight of time—that version of time defined by the "Señores del Tiempo Horizontal" ("Lords of Horizontal Time"). However, Julio Abril manages to overcome the imposed logic and gains access to the whole in a state of visionary hypersensitivity. Through dreams and latent myths, he wishes to discover a language that will expose the enigma that separated the external and internal signs of reality. This language, or a magic spell, "expresses the essence of all things."[41] Along with the mythical Casandra, Julio Abril rebels against the social conformity and scientific progress endorsed by the oligopolies in order to rescue human freedom.[42] The novel proceeds to describe the present world's crisis as a process of competitiveness fostered by two leading adversaries[43] and a system of propaganda. The system creates an inescapable determinism and a manipulative language that serves as its foundation. After Julio Abril's hidden energy is revealed, the duopoly attempts to channel and use this energy as a coercive force to foster its own interests. By refusing to collaborate, Julio Abril is sacrificed, but also liberated amidst a universe where the only language used to express the essence of the whole is revealed: "From the mysterious dot to the smallest level of creation, all things are dressed up with other things."[44]

In his novel *En la Gran Manzana (In the Big Apple)*,[45] Diéquez creates a futuristic vision of the world dominated by the manipulation of images and language that recalls the spirit of George Orwell's novel *1984*.[46] In this novel, uncertainty pervades the destiny of the human beings who live in a technocratic, global dictatorship, particularly the defenseless marginal groups, the unemployed, and immigrants who impede smooth competitiveness among large corporations. The end of the novel shows the vulnerability of the human being, the current lack of liberty and imagination, the unquestionable endorsement of technological advances, and the widespread acceptance of determinism rationalized by physical laws. The issues raised in this novel can be easily extrapolated to the problems of today's society. Moreover, Diéguez's work contains the elements that quantum aesthetics offers to solve these problems: mystery plus difference, creative freedom, democracy, individuality, respect for diversity, the search for beauty, and preference for the unique rather than the common. These are precisely the existential considerations that are suppressed by institutional power in Diéguez's scathing novel.

Quantum Poetry

Quantum influences are also apparent in poetry. Within a quantum framework, a poem can be understood as the materialization of one of the many semantic possibilities that fluctuate within a pretext of infinite potential. The quantum preoccupation with the interpenetration of the whole and the parts can also be seen in poetry. For example, in *Quantum Poetics: Yeats, Pound, Eliot, and the Science of Modernism*, Daniel Albright wonders: "Do we understand the whole text by understanding each of its parts, or do we understand each part of the text by understanding the whole?"[47] Indeed, the whole influences how the parts of a text are understood, as literary critics have recently asserted, but clearly the whole is also shaped by the parts. In this sense, the concept of complementarity might establish a bidirectional and simultaneous correspondence of meaning between the whole and the parts. This meaning, however, depends not only on the observer, but also on the quantum state of the text when it is observed. Furthermore, the poetic discourse reflects the principle of mystery plus difference articulated by quantum aesthetics, for both poetry and quantum theory have originality and personality as their central concerns.

For Rafael Guillén, poetry investigates the hidden angles of reality and represents a linguistic convergence of the conscious and the ineffable. When such integration occurs, words obstruct consciousness and pure intuition becomes the fundamental form of understanding. One of Guillén's recurrent themes is his decisiveness in surpassing the conventional limits of reality. In a number of his poems, he not only shows the deceitfulness of the senses, but also how much power we grant to them. In his book *Límites* (*Limits*),[48] he expresses complex ideas based on form, method, case, appearance, climate, and gestures. However, he never expresses exactly what he perceives. Regardless, there are unexpected occasions when lucidity breaks the boundaries of perception and allows the poet to enter momentarily the hologram of totality. When this experience occurs, he feels as if his body is a stranger, for it is constrained by space and time. At this moment, everything stops and Guillén is able to grasp a pulse of existence: "Being for an instant. Being immersed in other things / that exist. Afterwards there is nothing. Afterwards, the universe / continues its spinning death in a vacuum. / But for a second, it stops and lives."[49]

On other occasions, Guillén explores the interaction between consciousness and matter, as well as the distortion produced by the senses. His objective is to affirm that reality is not fixed, but rather probable and that it consists of potentiality and uncertainty. Once this correspondence—by definition, not subject to the laws of physics—occurs between consciousness and matter through observation, the mechanism that governs the totality of the physical universe breaks down and the quantum system acquires one of its possible states: "Only shadows and noises / live in the mansion. Who makes them? / The first thing is to know—and this is not possible—/ if they have life or they are given life / by our presence; if their being generates wonder or takes us out of our ignorance. / Above the eyes, in the most sensitive part, / sharpness gravitates."[50]

Another theory derived from quantum propositions, which deals with the ramification of universes,[51] is incorporated by Guillén into a number of his poems. In "Texto para Debussy" ("Text for Debussy"), for example, images wait for "the eyes that make them / reality."[52] Any consciousness of this event generates anxiety, since the materialization or nonmaterialization of the different possibilities depends on a choice that renders others unrealizable. However, the poet perceives intuitively that nonrealized possibilities remain in other universes: "[W]hat could have happened if one act / had not meant the exclusion of / all the others in just one moment!"[53] In "Vieja Fotografía en Sepia" ("Old Photograph in Off-White"), a parallel universe arises from the gesture of a young girl who is captured in a photograph. This gesture "in a different dimension / continued its natural / course."[54] In the poem "El Muro de Cristal"

("The Glass Wall"), Guillén describes the cries from another dimension of a nonexistent Berlin where Nazism has triumphed.[55]

Contact with other people also generates virtualizations that may not be actualized in the present universe, but may be realized in other worlds. An example of this is found in the poem "Signos en el Polvo" ("Signs in the Dust"),[56] in which the poet crosses paths with someone who is not aware of him. However, this event has an effect, perhaps ciphered, that never becomes real in the poet's world. In "Algo Sucede" ("Something Happens"),[57] the haphazard touch of a woman who does not see him makes Guillén perceive that something is happening somewhere else, even though he cannot explain this event. As Alistair Rae explains, "a ramification can occur whenever an interaction between two components of the universe happens."[58] A new dimension unfolds that no longer converges with the one in which the poet lives. Thus, for the poet, this new world is lost: "however, something / happens. I don't know where or how, / something huge happens that remains / written somewhere in everlasting and illegible print."[59]

This theory of ramification also maintains that the state of any given universe is the result of its past and the cause of the future of any universe that may diverge from it. In "Temporal" ("Storm"),[60] Guillén articulates a holistic notion of time—with limits that are relative or that do not exist—in which present events modify not only the future, but also the past. This notion of time can be seen in his poem "El Ayer es Mañana" ("Yesterday Is Tomorrow"): "Each moment has / two doors and two keys; / one of them doesn't close / and the other one doesn't open."[61]

Guillén combines a logical and intuitive reflection on matter and time with a mobile reference point. Accordingly, the myriad of countries in which the poems in *Los Estados Tansparentes* (*Transparent States*)[62] were written serves to enrich the imagery and perception that are necessary for a revolutionary change in our thinking. Reality is no longer a static physical world, but rather consists of the different states into which this world evolves in each moment. Reality no longer *is* but rather *becomes*.

Julia Uceda's poetry appears magnetized by the typical environment of estrangement that arises from the poet's confrontation with both everyday life and the conventional understanding of space and time: "I always was a stranger.... I am still a stranger."[63] Uceda defies language in order to dissolve the barriers that constrain human beings so that these persons can be revealed. Her penetration into the various and harmonious possibilities of being within reality gives to her poetry a mysterious profile. Based on everyday life, memories, dreams, external and internal time, alchemistic mysticism, and oriental philosophy, Uceda's poems illuminate the continuous exchange between the conscious and the unconscious. Her purpose is to reconcile the material and the immaterial, the spiritual and the worldly, day and night, the temporal and the eternal, life and death, and the rational and the irrational.

In Uceda's poetry, the essential importance of memories and dreams, as psychological realities that appear spontaneously, is emphasized. The poet does not choose her memories or dreams, but rather they choose her. Although they may arise in a whimsical, fragmented, or illogical manner, according to Uceda, memories distinguish people by the unique ways in which persons remember. Accordingly, the poet speaks *from* herself and not *of* herself. Additionally, while influenced by Jung, Uceda demonstrates that a new understanding emerges from dreams that allows for the recovery of earlier times and situations. Furthermore, dreams order the chaos that immediate reality can produce.

Light, Uceda's main emblem, emanates from every angle of her poetry as a holistic principle without beginning or end. Accordingly, light unchains the reciprocal transference between the symbolic, alchemistic, mystic, and personal planes that penetrate into the unknown and reveal traces and hints of the human being: "In this never-ending day / in which vocabularies are based on light, / and seeing is considered sufficient, / why name things?"[64]

In Manuel Mantero's books of poems, the confrontation between human beings and their understanding of reality is shown in order to express, ultimately, an interpretation of being and existence from a metaphysical perspective. Mantero's self unfolds continuously, thereby creating multiple selves that explore love, death, the everyday life, the individual, the collective, the exile, the occult, and the paranormal. The poet, therefore, becomes the dynamic center where Mantero the lover, the poet, the social being, the existential, the Spaniard, the exiled, the apparent, the authentic, the dreamlike, and the hermetic (mythic and mystic) converge and coexist. He reaffirms this pluralistic image in his poetry and in himself when he writes: "I am not one, I am many."[65] This pluralistic "I" is projected, ultimately, toward a holistic image of the world that threatens any attempt to restrict reality.

In Mantero's poetic writings, mythologies, ancient Eastern and Western religions, psychic and mystic traditions, philosophical approaches, and his own perceptions and intuitions are combined in order to maximize all the dimensions (sexual, literary, symbolic, natural, extranatural, paranormal, and transcendent) of creation. Through this creation, Mantero wants to find the thread that connects all the great doctrines and mysteries to the social and cultural dilemmas that surround the human being nowadays.

Mantero's verses represent the poet's active attempt to find the harmonic unity that conceals the multiplicity of reality. According to Mantero, the destiny of the human being is to reconcile the universal forces within us and understand, through inner contemplation, the fact that eternity—the final frontier of human existence—is the symbol of life. Thus, Mantero extols the vital experience of love and sex, as well as overcoming the world of appearances in order to reveal alterity: "The first time I saw you nude / I knew that a woman was more than just a precious façade. / She is nightly air with a transparent ceiling, / she is mud bequeathing, / she is a flame in a tiger, / she is the rain that soaks us wet: spring of existence. / And an apple overwhelmed nothingness."[66]

In José Hierro's poetic discourse, individuality on the other hand is perceived as a multiple and harmonious receptacle of events interpenetrated by subjectivity and objectivity, time and space, permanence and brevity, pain and joy, and expression and ineffability. On the other hand, Hierro alters the spatiotemporal conventions and enters the hologram of totality: "I am a traveler who came / from a different time / but I do not know whether this time is past or future."[67]

Hierro's verses use the dream and vigil as compelling expressions of both the conscious (reflection and poetic elaboration) and unconscious (and the states that trigger entry to this realm: sleep, enthrallment, and hallucination) worlds.[68] Through these processes, the poet receives fragments of humanity's heritage—forces and instincts accumulated in the collective unconscious throughout history—that have been translated into the language of myths and symbols. The integration of both the conscious and unconscious worlds into a harmonious unit allows human beings to explore a forgotten, but active domain that encloses the legacy of our common nature: "I add to the ocean. I rescue / gusts of creatures, gusts of human beat, creatures of the rain, / revived, innocently nocturnal, gusts. / Gusts, gusts, gusts / carved in mere shade."[69]

In addition, Hierro's poems address the quantum idea that objective reality is the result of the mental creations of observers, as they interpret certain patterns of vibration and frequency that comprise the universe. Interferences in the flow of frequencies provoke the transformation of atemporality (totality) into transitory, ordered, or materialized states (reality zones): "So much light, such open paths."[70]

For Ana María Fagundo, poetry is the mechanism that articulates her *self* and *the other* within a natural interpenetration of the mundane and the mysterious, the explicated and implicated orders.[71] Through her verses, Fagundo searches for integration between transitoriness and permanence so that an atemporal framework that promotes a holistic understanding of "self-being" ("serse") is established out of apparent fragmentation. In addition, the combination of consciousness and unconsciousness produces a domain of freedom where the poet finds her life and identity, that is, a sense of wonder when confronting the boundless nature of reality: "I, strange, upright being."[72]

The poem also becomes an existential foundation for Fagundo, because vital and creative processes are closely united to create a universe that is very personal. In this domain, Fagundo's voice feels the light of being and anticipates a possible transcendence, a "struggle of being, / of affirmation, mud inflamed through singing."[73] However, the dominant characteristic in Fagundo's books may not be that poetry and the poet are unique and irreducible, but rather that they are connected in order to produce art and conceptualize the world. Accordingly, Fagundo's testimony is revealed in her double condition as both woman and poet, a condition that resembles light's duality as wave and particle. So much so that the consideration of one or the other domain, by itself, does not do justice to her holistic nature: "This being of my being is now / infinite puzzlement, / pure certainty."[74]

In his poetry, Fernando de Villena characterizes consciousness as the illuminating hinge between planes of reality—the "I" and the other, and the individual and collective destiny of a human being in crisis. The poet promotes the extraordinary implications of an explicated order that is surrounded by an implicated order ("a different order without limits").[75] Villena's verses search for beauty through the use of baroque techniques of expression and the revaluation of ancestral legacies conveyed through myths. Thus, an arrangement between subjectivism and cultural heritage is produced that transforms the immediate reality into a totality. Women, central components in Villena's poetry, confer transcendence to the world. Through women, the poet fuses with primordial forces and discovers an intimate universe: "The world and its vertigo tire me. / I want you alone to create a world."[76]

For Francisco Plata, poetry is the means to express his discontent, suffocation, doubt, amazement, and rebelliousness. His verses reveal his perception of an order beyond the immediate that escapes formulation: "I am aware that the most important things remain unsaid; also I know that I will never be able to express many of the ethic and aesthetic keys."[77] Plata searches for his difference and mystery through knowledge, intimacy, symbolic naturalness, and the passion to discover what is extraordinary in our memories and our contact with others, as well as in the constant interpenetration of life and death. The invisible and the boundless rise symbolically against reductionistic ideas and question the nature of reality: " 'cause I am afraid that City has no end, / that from sidewalk to sidewalk there is as much distance / as there are possibilities to cross such distance—infinite. / City has no end. Does anything end?"[78]

In 1998, an excellent anthology prepared by de Villena was published, *La Poesía que Llega: Jóvenes Poetas Españoles* (*Coming Poetry: Young Spanish Poets*), in which the voices of Pedro José Vizoso, Rosario de Gorostegui, José Enrique Salcedo, Juan Carlos Martínez Manzano, Josefa Carmen Fernández Garzón, David Delfín, Miguel Ángel Contreras, Julio César Jiménez, José Luis Abraham, and Francisco Plata illustrate an original turn in present-day Spanish poetry. According to Villena, each of these poets represents "a fresh understanding of reality (their reality), a new and different vision."[79] These poets understand art as a spiritual and expressive adventure that both unravels the various levels of reality and strives to achieve a holistic vision. Morales classifies these authors as quantum poets because of their detachment from shallow realism, their holistic conception of time—which contains past, present, and future—their use of complex logic, and their belief in an implicate order, a subjective reality, and multiple universes.[80]

CONCLUSION

Literature is enriched in various ways by quantum ideas. This novel way of understanding the world and artistic creation demands reflexivity, from both writers and readers, so that they meet in a common communicative sphere where the potential text can reveal its explicit, albeit not necessarily exclusive, meaning.[81] The dynamic nature of the literary (or cosmic) universe requires the defeat of mechanistic and shallow realism and the establishment of a creative consciousness: "[W]e see the universe in the way it is because we exist."[82] The complexity of human experience demands a holistic vision in which (1) an underlying and indivisible totality, (2) an integrated matrix of internal and external realities, and (3) the interaction between the local energy of the parts and the active information of the whole are revealed. This is what the works of the poets and writers discussed in this chapter show. These authors share the quantum maxim that instructs them to reveal their mystery plus difference. They base their artistic freedom on intersubjectivity and imagination, in order to trigger linguistic amazement, narrative novelty, the manifestation of the extraordinary, the desire for self-exploration, and the search for beauty.

NOTES

1. Translated into English by Maritza Flores. Because of copyright issues, the original Spanish text of the poems discussed in this chapter have been omitted. Only Flores's translations of these excerpts have been included.

2. Although it is generally accepted that Newton strengthened the infinitesimal calculus of magnetism and electricity, the crucial contribution made by Gottfried Wilhelm Leibniz (1646–1716) cannot be ignored.

3. Years before the wave function became a central issue in quantum mechanics, Einstein proposed that atomic processes should be understood in probabilistic terms. However, Einstein was not seduced by quantum uncertainty and actually questioned this principle by affirming that God does not play dice with the universe. Although Einstein influenced decisively the development of quantum theory, he never entirely accepted this viewpoint. Rather than erroneous, he considered that quantum theory was incomplete.

4. Heisenberg challenged causality, one of the most revered principles of classical physics. By using statistical methods to describe quantum events, Heisenberg showed that subjectivity cannot be ignored. Furthermore, quantum physics affirms that a phenomenon cannot be observed without influencing the way it is revealed. Planck, Einstein, and Schrödinger opposed this characterization by believing in the underlying continuity of the physical world, along with the inexorability of causality in both the microscopic and the visible levels of reality. Heisenberg, however, refuted these ideas in two ways. First, he demonstrated that there is no evidence of "hidden variables" or invisible causal connections. And second, he asserted that not a single experiment has proven that the principle of uncertainty is incomplete. See Menas Kafatos, and Robert Nadeau, *The Conscious Universe: Part and Whole in Modern Physical Theory*. (New York: Springer-Verlag, 1990), Chapter 3.

5. Kafatos and Nadeau debate Heidegger's philosophy, Eastern metaphysics, Noam Chomsky's tranformational grammar, Ferdinand de Saussare's structuralism, Roman Jacobson, Claude Lévi-Strauss, A. J. Greimas, Roland Barthes, Michel Focault's postmodernism, Jacques Lacan, Jacques Derrida, imaginary numbers, geometry and algebra, neuroscience, biology, and psychology. Ibid.

6. Bruce Wilshire writes: "The very meaning of 'objective reality' must change considerably if we are to deliver ourselves from the abstraction of boxes of seventeenth-century physics and psycho-physical dualism, and understand how our own subjectivity, our own points of view on things and our relativity, are as 'objectively real' as anything else (just a little bit more weird and intriguing!). We ourselves are out there in the world, but not merely as objects viewed through a Cartesian abstraction called 'objectivity,' as if we squinted at ourselves through an optical instrument, but as beings participating in the archaic energies of others, Earth and the cosmos." (Bruce Wilshire, *The Moral Collapse of the University: Professionalism, Purity, and Alienation* [Albany: SUNY Press, 1990], 236).

7. Dick Richards, *Artful Work: Awakening Joy, Meaning, and Commitment in the Workplace* (San Francisco: Berrett-Koehler, 1995).

8. Margaret Wheatley, *Leadership and the New Science: Learning about Organization in an Orderly Universe* (Albany: SUNY Press, 1992).

9. Deepak Chopra, *Quantum Healing: Exploring the Frontiers of Mind/Body Medicine* (New York: Bantam, 1990).

10. In *El Cadáver de Balzac*, Gregorio Morales suggests that the relationships between Jungian theories and quantum ideas appear in a letter dated on February 27, 1953, that Jung sent to Pauli. In this letter, we can identify the following connections between quantum physics and psychology: "Quantum Physics is similar to a psychology that deals with the process of individuation and the unconscious; the impossibility of measuring a particle's position and speed at the same time is similar to the duality of scientific and intuitive knowledge; the impossibility of breaking an experiment into pieces without altering the observed phenomenon substantially is similar to the interpenetration of both the conscious and unconscious sides of the human being; the observer's effect on the observation is similar to the transformation of consciousness and the unconscious whenever consciousness is expanded; being comprised of a nucleus and outer layers, the atom is similar to the human personality that is comprised of a 'nucleus' (*or me*) and an 'ego' (*or I*)" ("La física cúantica a la psicología del proceso de individuación y del inconsciente en general, la imposibilidad de medir al mismo tiempo la posición y la velocidad de una partícula (ley de complementariedad) a la antítesis que forman el pensamiento científico y el saber intuitivo; la imposibilidad del subdividir el dispositivo experimental, sin modificar esencialmente el fenómeno, a la unidad golbal en el hombre de lo consciente y lo inconsciente; la modificación de la observación por el observador a la modificación de la consciencia y el inconsciente en cada adquisición de conciencia; el átomo, formado por núcleo y corteza, a la personalidad humana, formada por 'núcleo' (*o sí mismo*) y 'ego' (*o yo.*)" Gregorio Morales, *El Cadáver de Balzac* (Alicante: Epígono, 1998), 107.

11. Flora Lewis also comments: "In political terms, this means a great deal of restraint in thinking that we know best what is good for other people, that we must try to save Central America from itself or change the system of rule in the Soviet Union. In our own society, it means making sure we broaden the opportunities for all Americans to contribute. Who knows which people or group may be the magic quark that can produce great new benefits or beauty?" (Flora Lewis, "The Quantum Mechanics of Politics," the *New York Times*, 6 November 1983. See also Chapter 9 in this volume.

12. For example, see Michael A. Burayidi, ed., *Multiculturalism in a Cross-National Perspective* (Lanham, MD: University Press of America, 1997).

13. José Luis Gómez-Martínez reveals: "The anthropic principle implies a departure from the modern concept of a 'cultural center' and placing 'structure' in the foreground of explanation. The anthropic center is dynamic, mobile, and subject to the continuous transformations characteristic of any axiological discourse. This center can only be conceived within the dynamic process of its own contextualization, including the activity of (re)defining this context" ("La antropocidad implica una abstracción del concepto de 'centro cultural' que aporta la modernidad . . . , para colocar en primer plano la 'estructura' misma. El centro antrópico es un centro dinámico, móvil, un centro sujeto a la continua transformación propia de todo discurso axiológico. Es un centro que sólo se concibe en el proceso dinámico de su contextualización y como núcleo de constante re-codificación de dicha contextualización.") José Luis Gómez-Martínez, *Más Allá de la Pos-Modernidad: El Discurso Antrópico y su Praxis en la Cultura Ibeoamericana* (Madrid: Mileto Ediciones, 1999), 27–28.

14. Jean Paul Sartre, *What Is Literature?* (New York: Washington Square, 1966).

15. Bertolt Brecht (1898–1956) illustrates this circumstance in his personal diary. His notes are difficult to read as well as understand, for he did not intend to publish these ideas. On March 17, 1942, he wrote: "[P]hilosophers get irritated by heisenberg's proposition, according to which points in space and points in time cannot be coordinated. even if this had identified a limit beyond which descriptive methods theoretically cannot be 'improved,' the philosophers would still be left with the question of the possibility of description, so that their proposition that nothing happens without cause would still stand. the physicists have overturned it by demonstrating its emptiness; they just abandon it. grounds that cannot be established theoretically are not grounds for them at all." The next day, he commented: "[I] like the world of physicists. men change it, and then it looks astonishing. we can appear as the gamblers we are, without our approximations, our to-the-best-of-our-abilities, our dependence on others, on the unknown, on things complete in themselves. so once again a variety of things can lead to success, more than just one path is open. oddly enough i feel more free in this world than in the old one." Bertolt Brecht, *Journals*, trans. Hugh Rorrison, ed. John Willet (New York: Routledge, 1993), 208–209.

16. García Viñó indicates the following characteristics of the relativistic and quantum novel: (1) themes such as the nonexistence of space, the inversion of time, or the existence of parallel universes are introduced; (2) heterodoxy is preferable to abiding by fixed rules; (3) the plot does not evolve chronologically but (4) takes place within a psychological rather than a physical space; (5) the characters embody ideas and events rather than human beings; (6) the novel aspires to be a work of art instead of an imitation of the world; (7) both author and reader, rather than the characters, live the events; (8) the world comprises events, rather than objects and people; (9) the plot does not follow a formal logic, but rather one created ad hoc by the author and the novel itself; (10) the events of the present influence and modify both the future and the past; (11) rather than the classical predetermined plot, quantum and relativistic novels present just one of an infinite number of possibilities; (12) the novel generates characters and events, rather than the other way around; (13) time and space are the main concerns; the relationship between future and past, and distance and size, reside in the consciousness of the characters, rather than in their "external" world; (14) the reader perceives time and space through character's consciousness; (15) the distortion of space-time is a rich source of estrangement and, thus, of aesthetic values; (16) past and future are as real and accessible as the present; dreams, memories, and desires are

as real as anything that occurs "outside" of the person; and what "should be" is as real as what it "actually is." See Manuel García Viñó, "La novela Relativista y Cuántica: Materiales para la Construcción de una Teoría Aplicable a Otras Artes," *Heterodoxia* 22, (1995). The following articles by García Viñó can also be consulted: "La novela y la Nueva Física," *Cuadernos del Sur* 6 July 1995, 4–5; and "Carácter Forzoso de una Novela Relativista y Cuántica," *Papel Literario*, 5 October 1997.

17. Gregorio Morales reveals the fundamental tenets of a global quantum aesthetic: "(1) The artist has the power to break away from the walls imposed by Newton. (2) Reality is not restricted to what is visible or knowable. (3) The implicate and invisible order is differentiated from the explicate and manifest order. (4) Matter and spirit are the same thing; they form a continuum. (5) The laws of causality become relative. In fact, we can even find an acausal relationship between the psychological and the physical realms—a synchronicity. (6) Nothing is distinct. Even the most distant corpuscles influence one another. (7) Both observer and observed are tied together. Consequently, after a few centuries in exile, the human being returns to inhabit the center of the universe. (8) Every human experience becomes a part of what we call morphogenic fields. These can be entered through intuition. (9) We can remain within the visible world, that is, the preceding tenets should not be taken as fetishes. (10) The task of every human being is individuation, that is, jettisoning our collective unconscious so that we can be 'ourselves.' This is translated into a primary concern for freedom and diversity. (11) Accordingly, we have to assume an ethics based on tolerance, anti-dognatism, and a strong defense of democracy. (12) We need to search for beauty. (13) We prefer the extraordinary rather than the common, eroticism rather than the physiology of sex, maturity rather than juvenility, individuality rather than collectivity, internal transformation rather than surgery, and consciousness rather than fluids. (14) All of these tenets can be summarized as follows: *Mystery plus difference*" ["1) Potestad del artista para franquear los límites de las apariencias más allá de las paredes newtonianas. 2) La realidad no sólo es lo visible ni lo cognoscible. 3) Diferenciación entre un orden plegado e invisible o otro manifiesto y palpable. 4) No concreción entre materia y espíritu. Ambos son la misma cosa. Forman un coninuum. 5) Relativización de las leyes de causa y efecto. Puede darse una relación acasual entre lo psíquico y lo físico, es decir, lo que se llama *una sincronía*. 6) No existe la separabilidad. Hasta los corpúsculos alejados por millones de años luz se influyen recíprocamente. 7) No diferenciación entre el observador y lo observado. Consiguientemente, el ser humano, después de siglos de destierro, vuelve a habitar en el centro del Universo. 8) Toda experiencia humana queda impresa en lo que se denominan *campos morfogenéticos*, accesibles de una manera intuitiva. 9) Voluntad de permanecer en el mundo visible, sin que las anteriores "licencias" puedan ser motivo de inflación ni de misticismo. 10) La tarea de todo ser humano es la *individuación*, es decir, el modo en que se libera del inconsciente colectivo para ser todo 'él mismo.' Lo que se traduce en una atención primordial a la libertad y a la diversidad. 11) Asunción, como consecuencia de lo anterior, de un plano ético, del que se desprende la tolerancia, el antidogmatismo y la decidida defensa de la democracia. 12) Búsqueda de la belleza. 13) Preferencia por lo extraordinario antes que por lo común; por el erotismo, antes que por la fisiologia del sexo; por la madurez, antes que por el juvenismo; por la conciencia antes que por los fluidos. 14) Todos los anteriores principios pueden resumirse en el sigiente: *Misterio más diferencia*."] Morales, *El Cadáver de Balzac*, 38–39.

18. In the *Quantum Aesthetics Group E-m@ilfesto*, the basic principles of a new conception of art are outlined: "1) The placement of mankind back into the role of active producer of the Universe. 2) The certitude that wo/men are the creators of their reality. 3) The integration of opposites into the whole. 4) The understanding of matter and soul as two aspects of the same magma, which can influence each other and produce, among other possibilities, the so called synchronicities, or significant chances. Furthermore, the possibility that matter could be intelligent: As Eddington says, the universe is constituted by mindful matter. 5) The belief that life is 'individuation.' That is, life is the process whereby each person brings his/her own singularity to the surface. Since every individuated person necessarily contributes to his/her community, individuation is thus radically opposed to *individualism* or *egotism*. 6) The conviction that every literary and artistic work ought to have a holographic essence if it is to be in tune with the universe—in which case, the smallest part always contains the whole. 7) Persons have the capability to penetrate the 'morphogenetic fields' (Sheldrake) where the destiny of humanity and the universe reside. 8) The quantum aesthetics group has a vital stake in the type of civilization and consciousness that emerges from the integration of Nature and the unconscious. 9) The understanding of the cosmos as a compact fluid in which, as the non-separability theory states, everything is interconnected. 10) The overcoming of the positivist theory of relativity, since there exists a speed faster than that of light. At least, in the polarization of subatomic particles, such a speed is possible." Quantum Aesthetics Group, *Quantum Aesthetics Group's E-m@ilfesto*, 20 October 2000 ‹http://teleline.terra.es/personal/lucschok/estetica/ emailfestoeng.htm›. The Web page of the group can be consulted at ‹http://www. quantum-aesthetics.com›.

19. This meeting took place in Valencia, Spain, in July 1995.

20. Morales, *El Cadáver de Balzac*, 189.

21. Manuel García Viñó. *El Puente de los Siglos* (Madrid: Ibérico Europea de Ediciones, 1986).

22. ("Todo cuanto habíamos construido juntos, tú, en este instante, con tu traición, con tu falsedad, tu cobardía lo acabas de destruir. Porque nuestros actos, Arturo, no sólo influyen y determinan nuestro futuro; nuestros actos presentes modifican nuestro pasado también.") Ibid., 22.

23. "The perishable and temporal world is not maya in itself, as Hinduism suggests. Besides that world, there exists another maya world—I've discovered it—comprised of pre- and post-reflections on the totality of the world" (*"El mundo temporal perecedero no es en sí mismo un maya, como piensa el hinduismo. Pero, aparte de él existe (yo lo he descubierto) un mundo-maya que consiste en el pre y post-espejamiento del acontecer total del mundo."*) Ibid., 86.

24. ("Nosotros hemos descubierto la droga . . . Dro-ga, amigo mío, no máquina; la droga capaz de desintegrar a un ser humano y volverlo a integrar en un punto previamente calculado del mundo-maya, que a su vez será, naturalmente, integrante del mundo temporal-perecedero, pero distinto, claro está . . . Ah, y, por lo que ahora podemos, pasado.") Ibid., 86.

25. ("Cada uno de vosotros, mortales es el centro de *un* Universo, su Universo, uno de los infinitos universos existentes. Todo lo imaginable es posible . . . Muchos de los acontecimientos, de las cosas de tu Universo sirven para otros morales que viven *simultáneamente*, al menos en cierta medida, a veces insignificante, a veces consider-

able, contigo; mortales cuyos universos se interfieren con el tuyo y que no son, al menos en esta medida, incompatibles con él, tu Universo.") Ibid., 182–183.

26. ("El muerto, el no-nacido, el resucitado, el vivo comprendió que estaba en el centro del mundo y que algo, en algún lugar, había recomenzado.") Ibid, 184.

27. Gregorio Morales, *La Cuarta Locura* (Barcelona: Grijalbo, 1989).

28. "Beauty may be no more than a model we all have imprinted deep within ourselves, a model we are compelled to pursue throughout our lives. But, like I said, Iria Clarós was not so much a model as she was a flame that set me on fire. Her contact was capable of awakening my mania, the one that Plato had called the *fourth madness*—the madness that makes us search incessantly for beauty. If we do not follow this impulse, as I hadn't for a long time, life loses its meaning. Something tears us apart from inside, as if carnivorous animals lived in our hearts and howled for their food. That's the way the fourth madness works. It is the longing for beauty. Once you've observed the beautiful things in the world, the desire grows in your soul to fly off and discover the essence of beauty" ("Tal vez la belleza no sea más que un modelo que todos llevamos profundamente impreso y al que nos vemos obligados a perseguir durante toda la vida. Pero, como he dicho, Iria Clarós, más que modelo, fue fuego que me incendió. El contacto con ella tuvo la virtud de despertar en mí la *loca manía*, la que Platón llamaba *cuarta locura*, la locura que ya siempre nos hace buscar la belleza. Y si no obedecemos este impulso, como me había ocurrido a mí durante tanto tiempo, la vida pierde sentido para nosotros. Algo nos desgarra interiormente, como un animal carnívoro alojado en el corazón, pidiendo a rugidos su alimento. Así actúa la *cuarta locura*. Es el ansia de belleza. Una vez contempladas las cosas bellas de este mundo, crece en el alma el deseo de tomar alas y emprender el vuelo para llegar a la esencia primera de toda hermosura.") Ibid., 33.

29. According to Carl G. Jung, the process of individuation requires the differentiation of what is one's own from what is common to others, in order to learn about oneself—as a unique and irreplaceable being. In this sense, individuation is a transformation whereby the human being reconciles the unconscious and conscious forces that surround him/her and attempts to control them. See Carl G. Jung, *Man and His Symbols* (London: Aldus, 1972). For Iria Clarós, her compulsions are ways to integrate her shadow. As psychoanalyst Kartus describes: "and in the mirror, what some people have called the *double* or the *shadow*—that is, the twin self or dark 'I' that everybody avoids and hates—shows up" ("y en el espejo aparecía lo que algunos han lllamado *el doble o al sombra*, es decir, el yo gemelo, el yo oscuro que uno evita vivir y repudia.") Morales, *La Cuarta Locura*, 266.

Another of Morales's novels *Él. Ella*—which can be read by starting from either *Ella* or *Él*—is "a meditation on the feminine and masculine, because the progressive ambiguity of both terms is one of the biggest crises that the next generations will face" ("una meditación sobre lo femenino y lo masculino, cuya progresiva ambigüedad constituye uno de los mayores naufragios con que se encuentran las nuevas generaciones.") Gregorio Morales, "Preámbulo," in *Él. Ella* (Alicante: Epígono, 1999).

The individuation of Silvia and David, assisted by the conciliatory experience of both the Western and Eastern worlds—incarnated in Iván and Rada—establishes cooperation between the sexes through knowledge and control of the latent forces of the animus/anima and the shadow.

30. ("danza más subterránea, una danza de la tinieblas, de la que habrían de surgir los aspectos más desconocidos del ser.") Morales, *La Cuarta Locura*, 138.

31. Kartus tells Alzu: "Do you know why, despite all of modern technology, scientists have not found a way to mold human behavior? Because human behavior extends in millions of directions that are impossible to understand at the same time. For any given person, anything can exist and not exist simultaneously, be up and down, straight and crooked. . . . That's why I tell you: Imagine! Your imagination will always be as real as the so-called reality that you are trying to discover" ("¿Sabe por qué con todos los adelantos modernos los científicos no han podido aún reglar el comportamiento humano? Porque ese comportamiento se expande hacia millones de direcciones, imposibles de aprehender todas a la vez. Para cualquier persona una cosa puede simultáneamente ser o no ser, estar arriba y abajo, del derecho o del revés. . . . Por eso le digo: ¡imagine! Su imaginación será siempre tan real como la pretendida realidad que busca.") Ibid., 95.

32. ("como única forma de llegar a lo desconocido.") Ibid.

33. A hologram is a tridimensional image produced when a laser is divided into two beams. After reflecting the photographed object, the first ray collides with the second and produces an interference that is imprinted on to a photographic plate. Apparently, the image has nothing to do with the photographed object until it is illuminated by another laser. Then, a tridimensional image of the object appears that can be seen from different angles. The real significance of a hologram is that, after being divided repeatedly, each fragment reproduces the whole original object. In order to explain this phenomenon, David Bohm suggests that all subatomic particles are interconnected and that the whole is the most precise approximation of reality. For Bohm, everyday reality is a type of illusion, similar to a holographic image, that conceals a deeper level of existence—an implicate order—that underpins the immediate world of the senses—the explicate order. The materialization of one aspect of reality or another depends on the unification of, and the observer's interaction with, both orders. According to Bohm, wholeness is characterized as a unit without fragments; that is, fragmentation is arbitrary and only possible in the explicate order. See David Bohm, *Wholeness and the Implicate Order*. (London: Routledge and Kegan Paul, 1980). Accordingly, access to the whole can be accomplished from any of the parts, once we reach an alternative state of consciousness—a holotropic state. See Michael Talbot, *The Holographic Universe* (New York: HarperCollins, 1991). René Lange argues in *La Cuarta Locura*: "there are some people that, with their mere presence, can reveal the true dimension of things. As if things, wrapped in the darkness of a maelstrom, had been waiting for a long time for the opportunity to spout and grow. By becoming friends with those people, and seeking their assistance, we can grasp the true essence of art" ("hay un tipo de personas que, con su sola presencia, poseen la facultad de hacer salir a la superficie la dimensión permanente de las cosas, como si éstas, envueltas en las tinieblas de una nebulosa, hubieran estado esperando por largo tiempo la ocasión de brotar de si mismas. Si obtenemos la amistad de una de estas personas, es posible llegar a través de ella a la auténtica esencia del arte.") Morales, *La Cuarta Locura*, 32.

34. ("¿Qué ha perseguido a lo largo de su vida, en su obra y en el vértigo que ésta le producía? Llegar a la intuición de la muerte. ¡Es justamente eso lo que reflejan sus desnudos! Usted ha ido tras los temblores de la muerte, que es lo mismo que ir tras la esencia de la belleza.") Ibid., 366.

35. Gregorio Morales, *El Pecado del Adivino* (Madrid: Modadori, 1992).

36. According to Edgar Morin, "complexity is a fabric made out of heterogeneous parts that are inescapably tied together; complexity involves the paradox of including unity and multiplicity at the same time" ("la complejidad es un tejido de constituyentes heterogéneos inseparablemente asociados; presenta la paradoja de lo uno y lo múltiple.") Edgar Morin, *Introducción al Pensamiento Complejo* (Barcelona: Gedisa, 1994), 23. He goes on to say that complexity "also comprises uncertainties, indeterminations, and random phenomena. In a way, complexity is always related to chance" ("comprende también incertidumbes, indeterminaciones, fenómenos aleatorios. En un sentido, la complejidad siempre está relacionada con el azar.") Ibid., 60. In terms of complementarity, science shows contradictory properties, such as light's dual nature as wave and particle (a combination of apparent opposites, no doubt) or the integration of matter and mind—which could account for synchronicities.

For Kafatos and Naudeau, "complementarity is the most fundamental ordering principle in the conscious constructions of all human realities." Kafatos and Nadeau, *Conscious Universe*, 128).

37. Regarding Ángela and Adrián, the narrator says the following: "I am sure that, at that time, both Ángela and Adrián recognized each other as remains of a shipwreck; a fortune-teller and a priestess, both of them passengers on a boat that had crashed against time and, after wandering around lost oceans, had arrived at a place where light, color, noise, bodies, and the delight of giving oneself to empty conversations are celebrated" ("Estoy seguro de que Ángela y él tuvieron que reconocerse en aquella ocación como los restos del naufragio del pasado que eran, un adivino y una sacerdotisa, pasajeros de una misma nave que se hubiera estrellado contra el tiempo y que tras vagar por perdidos mares, arribara a este enclave que celebra la luz, el color, el ruido, los cuerpos, la embriaguez de entregarse a vacuas conversaciones.") Morales, *El Pecado del Adivino*, 22.

38. According to Jung, synchronicities are acausal relationships between the physical and psychological worlds. The importance of this concept resulted in Jung and Pauli coauthoring *The Interpretation and Nature of the Psyche* (New York: Pantheon, 1955). In Morales's novel, Adrián experiences a synchronicity when he wishes for Sister Ángela to appear, and "sure enough, not half an hour had past when he saw the reflection of her glasses and, right away, the white cape that waved on her back" ("en effecto, no había transcurrido media hora cuando de pronto atisbó el reflejo de sus gafas y, en seguida, la capa blanca que ondeaba a su espalda.") Morales, *El Pecado del Adivino*, 164.

39. ("ser el Universo y no una ínfima partícula de él.") Ibid., 239.

40. Miguel Ángel Diéguez, *Los Días del Duopolio* (Madrid: Libertarias/Prodhufi, S.A., 1989).

41. ("para expresar la esencia de todas las cosas.") Ibid., 330.

42. Casandra's ancestral words reveal a mental universe that is compatible with quantum principles: "The cosmos is an ocean of fermenting ideas. The universe is mental. Some of these ideas were realized in here, lived for centuries, and were altered constantly. The 'sign' meant a revolution. However, this uprising is being tamed, controlled, and destroyed in all of its manifestations, and thus your species is walking toward extinction; a hecatomb, a collective frenzy, persons dragging themselves around devouring the last residues of a planet that can be found. . . . 'They' know it. They are searching for the boiling ocean that stirs their burned out ideas. They will

never find it. Their brains are blocked. They have not been able to place themselves within levels of imagination that could help them break away from their voluntary imprisonment. They did not understand the past, and surely will not understand the future. Substituting gods or perfecting productive means is not an authentic answer. Actually, they hate freedom because they confuse it for their religious or scientific norms" ("El cosmos es un mar de ideas en fermentación. El universo es mental. Parte de esas ideas se corporizaron aquí y se prodigaron durante siglos en imaginativas mutaciones. La sublevación fue el signo. Pero la sublevación está siendo atrapada, controlada, destruida en todas sus manifestaciones, por lo que tu especie camina a la extinción. La hecatombe, el delirio colectivo, el arrastrarse devorando los últimos residuos que puedan extraer del planeta . . . 'Ellos' lo saben. Buscan el mar en ebullición que renueve sus ideas quemadas. Nunca lo encontrarán. Sus cerebros permanecen embotados. No han podido situarse en planos imaginativos que les ayuden a romper los espacios en que se han recluido. No entendieron el pasado, no entenderán el futuro. Sustituir los dioses o perfeccionar los métodos productivos no es el camino auténtico. Odian, en realidad, la liberación porque la confunden con sus normas sagradas o científicas.") Ibid., 294.

43. "Behind this apparent rivalry exists what they call the 'Alliance Ark.' In reality, the 'Ark' is a private agreement between two leaders that allows for the creation of an atmosphere of war that benefits both of them. . . . That 'Ark' . . . generates something fundamental: the rivalry between the two groups. For the rest of the bosses and servers, the 'Ark' is an unquestionable dogma that allows them to live, and hate and fight one another. This dogma has a goal that is both diabolical and simple: dividing the world into two belligerent camps. In a shallow but efficient Manichean way, persons are either good or evil depending on the side to which they are assigned. Competitiveness between the two sides produces something that is vital for both of them: PRODUCTIVITY. That is the key tenet of the duopoly" ("Tras la aparente rivalidad existe lo que llaman el 'Arca de la Alianza.' En realidad es un acuerdo privado entre ambos líderes que permite crear la atmósfera de guerra que a ellos les conviene. . . . Ese 'Arca' . . . genera algo fundamental: la rivalidad entre los dos grupos. Para los demás dirigentes y servidores en general constituye un dogma indiscutible que les permite vivir, odiarse, combatirse. Este dogma tiene una finalidad tan diabólica como simple: dividir el escenario en dos bloques beligerantes. El mal y el bien se aplican según el bloque al que pertenezcas, con un maniqueísmo ramplón pero eficiente. La competencia de los bloques crea algo vital para ambos: PRODUCTIVIDAD. Ese es el centro madre del duopolio.") Ibid., 285–286.

44. ("A partir del misterioso punto hasta el más ínfimo grado de la creación, todo sirve de vestidura a alguna otra cosa . . .") Ibid., 331.

45. Miguel Ángel Diéguez, *En la Gran Manzana* (Alicante: Epígono, 1997).

46. When mentioning the new image-synthesizers, Rojas explains: "Your images are child's play compared to what the synthesizer can do. This device is capable of creating and storing all kinds of images, so events that *never happened* can take place. That allows us to distribute through the mass media—essentially television—predesigned events that are compatible with our interest. In this way, the current model is inverted. We used to need an event—sometimes we provoked it with a lot of effort. . . . Then we manipulated it. Now, any event can be produced in a lab and appear as real, and does not need any posterior manipulation. An election can be won through a series of events that has been scientifically programmed. We can move

so-called public opinion to support A or B—and give them no possible alternative. People can be compelled to buy or refuse certain products" ("Tus imágenes son un juego de niños en comparación con lo que desarrollará el sintetizador. Es capaz de crear y almacenar todo tipo de imágenes para producir acontecimientos que *nunca han ocurrido*. Ello permite difundir a través de los medios de comunicación— televisión esencialmente—sucesos previamente diseñados con el fin de conseguir lo que a nuestro sistema convenga. De este modo se invierte el método actual, que exigía un suceso— provocado a veces, pero mediante un esfuerzo considerable . . .—y su manipulación posterior. Ahora el suceso se creará en los laboratorios, surgirá como real y no requerirá ningún tipo de manipulación posterior. Se pueden ganar unas elecciones con una serie de acontecimientos científicamente programados; se puede conducir a la llamada opinión pública a aceptar unas normas tendentes a A o B, sin salida posible; se puede obligar a la gente a comprar o rechazar ciertos productos.") Ibid., 192.

47. Daniel Albright, *Quantum Poetics: Yeats, Pound, Eliot, and the Science of Modernism* (Cambridge: Cambridge University Press, 1997), 5. Although he acknowledges how controversial his ideas may be, Albright uses "scientific metaphors" to write about the modernist poetry of W. B. Yeats, Ezra Pound, and T. S. Eliot. For Albright, this poetry is based on a pattern of waves and particles that can be considered quantum.

48. Rafael Guillén, *Límites* (Barcelona: Colección "El Bardo," 1971).

49. See the original Spanish version in Rafael Guillén, "Ser un Instante," *Límites*, 37.

50. See the original Spanish version in Rafael Guillén, "Introducción al Misterio," in *Límites*, 15. Consciousness, as a quantum mechanism of observation, enhances the subjective character of reality. Some scientists have suggested that consciousness is not only possessed by human beings, but, in some degree, also by other animals and inanimate objects. This idea is insinuated in several poems by Guillén. In "Las Raíces" ("Roots"), material things recognize human beings: "a picture, or a book, / they organize themselves when they feel we are close." See the original Spanish version in Rafael Guillén, *Los Alrededores del Tiempo: Antología* (Granada: Antonio Ubago, 1988), 121–122.

Discovering consciousness in things also appears in *Los Estados Transparentes*. For example, Guillén writes about the palpitations of marble (Rafael Guillén, "Ascensión del mármol," in *Los Estados Transparentes: Nueva Versión Revisada y Ampliada* [Granada: Pre-Textos, 1998], 136–137), and rocks ("Piedras para una Catedral de Monet," ibid., 52–54). He also writes about taxis that "caress their beaten-up / upholstery" at the end of the day (see the original Spanish version in ("De la Materia de los Taxis," ibid., 125–127), and about objects that "seethe with the tragedy / of survival" in an antique store (see the Spanish version in "Pasos en la Tienda de Antigüedades." ibid., 64–66.)

51. The consequences of this theory are revolutionary. Alastair Rae explains that every time that a quantum observation is produced, the universe is ramified in as many directions as an observation makes possible. Accordingly, this theory postulates the existence of a myriad of universes that, by definition, either occupy the same locale—even though they cannot converge except in very special and precise circumstances—or reside in different dimensions of space. See Alastair Rae, *Quantum Physics: Illusion or Reality?* New York: Cambridge University Press, 1986), 75–83.

52. See the original Spanish version in Rafael Guillén, "Texto para Debussy," in *Los Estados Transparentes*, 100–102.
53. See the original Spanish version in Rafael Guillén, "La configuración de lo perdido," in *Límites*, 51.
54. See the original Spanish version in Rafael Guillén, "Vieja Fotografía en Sepia," in *Los Estados Transparentes*, 57–58.
55. Rafael Guillén, "El Muro de Cristal," *Los Estados Transparentes*, 180–181.
56. Rafael Guillén, "Signos en el Polvo," *Límites*, 28.
57. Rafael Guillén, "Algo Sucede," *Límites*, 64–65.
58. Rae, *Quantum Physics*, 82.
59. See the original Spanish version in Rafael Guillén, "Algo Sucede," 64–65.
60. Rafael Guillén, "Temporal," in *Los Alrededores del Tiempo: Antología* (Granada: Antonio Ubago, 1988).
61. See the original Spanish version in Rafael Guillén, "El Ayer es Mañana," in *Los Alrededores del Tiempo*, 180.
62. Guillén, *Los Estados Tansparentes*.
63. See the original Spanish version in Julia Uceda, "La extraña," in *Mariposa en Cenizas* (Arcos de la Frontera: Alcaraván 1959), 36.
64. See the original Spanish version in Julia Uceda, "Campanas en Sansueña," in *Poesía* (El Ferrol: Sociedad de Cultura Valle-Inclán, 1991), 198.
65. See the original Spanish version in Manuel Mantero, "Intento de dos Justificaciones," *Ya Quiere Amanecer*. In *Como Llama en el Diamante. Poesías completas, vol. 3* (Sevilla: Colección Literaria, 1996), 59.
66. See the original Spanish version in Manuel Mantero, "Exculpa de la Manzana," *Ya Quiere Amanecer*. In *Como Llama en el Diamante, vol. 3*, 25.
67. See the original Spanish version in José Hierro, "Alma Mahler Hotel," in *Cuaderno de Nueva York*, 9 ed (Madrid: Hiperión, 1999), 52.
68. For Hierro, the mechanisms capable of inducing a holotropic state of consciousness are raptures, hallucinations, and dreams. Through these means, we can enter the holographic labyrinth that connects all the aspects of existence and reality. That is, we can reach a totality without fissures, a vast ocean of energy where space and time are undifferentiated and manifest themselves in a precise place in the past or future. See Bohm, *Wholeness and the Implicate Order*. There is a close association between these alternative states of consciousness and the unconscious. Through dreams or visions, the human being can penetrate the hologram created by the accumulation of humanity's psychological states and experiences. That is, from a fragment—from the individual—the human can reach the universal.
69. See the original Spanish version in José Hierro, "Un Continente Olvidado," in *Cuaderno de Nueva York*, 81.
70. (¡Tanta luz, tan abiertos caminos!") José Hierro, "Alucinación," in *Alegría* (Madrid: Ediciones Torremozas, 1991), 17.
71. See Bohm, *Wholeness and the Implicate Order*.
72. ("Yo, extraño ser enhiesto.") Ana María Fagundo, "Flor de Jara Junto al Camino," in *Trasterrado Marzo* (Sevilla: Colección de Poesía "Angaro," 1999), 83.
73. See the original Spanish version in Ana María Fagundo, "Amando," in *Obra poética 1965–1990* (Madrid: Ediciones Endimión, 1990), 307.
74. See the original Spanish version in Ana María Fagundo, "Ser en Mí," in *Trasterrado Marzo*, 27.

75. See the original Spanish version in Fernando de Villena, "Cautivo," in *Poesía (1980–1990)* (Granada: Ediciones A. Ubago, 1993), 112.
76. See the original Spanish version in Fernando de Villena, "Cuerpo y alma," in *Poesía* 295.
77. See the original Spanish version in Fernando de Villena, ed., *La Poesía que Llega: Jóvenes Poetas Españoles* (Madrid: Huerga y Fierro, 1998), 107.
78. See the original Spanish version in Francsico Plata, "Cuidad de México," in *The Modern-Post* 5 (Spring Semester 2000), 2.
79. ("una visión de la realidad (o de su realidad) plena de frescura, una visión nueva y distinta.") Villena, *La Poesía que Llega, 11*.
80. Gregorio Morales, "Poetas Cuánticos," *El Faro: Cultura*, 19 June 1998, 26.
81. See Mihaela Dvorac's discussion on Julia Kristeva's pheno- and genotext in Chapter 2 of this volume.
82. Stephen Hawking *Historia del Tiempo: Del Big Bang a los Agujeros Negros. ¿Cuál es la Naturaleza del Tiempo y del Universo?* (Barcelona: RBA Editores, 1993), 166. English version: Stephen Hawking, *A Brief History of Time: From the Big Bang to Black Holes* (New York: Bantam, 1988).

CHAPTER 4

Aesthetics as Creation and Transformation of World Awareness

Algis Mickunas

INTRODUCTION

Critics have said that art has many meanings, and indeed may even verge on meaninglessness. It may represent reality, evoke morality, intimate higher entities and lower images, express human feelings, and trace social relationships. Art may support the interests of the ruling classes, or be a consolation to the oppressed. Yet, across all such meanings, there is a common prejudgment. Art has value because it has an inherent truth. It tells a story about something, depicts something, reveals, manifests, and even hides things. The value of art resides in its metaphysical or ontological referent. Therefore, the principles of logical discourse that make judgments about these theoretical conceptions are equally valid to make assessments about art. In most cases, metaphysical and ontological referents are used as criteria to judge whether some human construct, such as a painting or a story, is or is not art.[1] Nonetheless, the same piece may be understood to be an artwork within one metaphysical or ontological framework, and trash, pornography, or immoral in another.

This chapter is not intent on resolving these controversies, but on showing that the different arts and artistic movements render ontological positions—worldviews—that are distinct. That is, this chapter intends to delve both into the nature of art and the ontological bases of reality, for art is both the mirror and raw material of reality. In order for me to show the differences among the various types of world awareness, first I need to show how these worlds—as space, time, and movement[2]—are a priori and the basis of everything else that exists in them. Thus, art cannot escape from or transcend the world, but only represent and help to create reality. That is, the phenomenal world is an a priori of everything that exists, including art. This chapter also tries to discuss the type of world, as a phenomenal a priori, that serves as the basis of quantum aesthetics, including the type of world quantum aesthetics helps to create.

The first section of the chapter thus deals with how the transcendent presupposes the phenomenal world. The next four sections focus on the bases of each of the different types of world awareness that exist—the worlds of identity, eternal return, perspective, and dynamic fields—and their connections to quantum aesthetics.

But before starting my discussion, however, a word of caution is needed. The different worlds to which this chapter refers need not necessarily be considered as evolutionary stages of world awareness.[3] From a quantum point of view, these worlds can be understood as possibilities of the implicate order[4] that only comes to fruition after a person, either consciously or unconsciously, thinks and thus constructs the world one way or another. Artists are very important in this process. That is, a specific world can become real only when people, especially artists, collapse its wave function.[5] Thus, the shifts from one world to the next represent quantum leaps, for they involve the collapse of a particular wave function.

FIRST PHILOSOPHY

Metaphysical thinking, in principle, attempts to derive and order the cosmos from a vantage point that posits timeless perfections—ideals. They are the source and condition—the transcendent X—of the world. This X is interpreted in various ways: as emanation, creation, unfolding, laws that order all things, and deduction. Common to all of them is that they underlie or lay beyond space, time, and movement, and hence the latter elements are only derivative phenomena. Nonpositional factors are regarded to exist "before" any being of the phenomenal world, and hence are independent of the realm of appearances. Here, first philosophies of whatever type confront a riddle. How can the X be regarded as beyond if it must assume a temporal location—before—and a spatial position—above or underneath—the phenomenal? If the metaphysical beyond and underneath are transtemporal and transspatial, and yet the condition for both, then this eternal domain assumes a position that grants the phenomenal world a status that exceeds the metaphysical. This is to say, in order to be the beginning or the transcendent and underlying metaphysical condition, the X must grant priority to space, time, and movement, for without them no one could understand that the phenomenal world is derived from X.[6] Immanuel Kant, for example, expresses this idea by showing how no judgment is possible without the necessary conditions of space, time, and movement.[7] Thus, the depth of the transcendent realm becomes coextensive with the depth of the phenomenal world. This transcendent consideration gains, loses, or is diminished with respect to the phenomenal world. The reverse also holds true: The depiction of this transcendent reality implicates the constitution of the phenomenal world.

No doubt, the variety of first philosophies suggests the protracted efforts to "explain" the phenomenal world of space, time, and movement by something more important, and, above all, more permanent, yet such explanations constantly call for phenomenal confirmations. Furthermore, while first philosophies may seem far-fetched, they are, nonetheless, adopted by all disciplines in their own unique way. Each one wants to offer an explanatory hypothesis for all the rest. Indeed, biologists defend their honor by arguing that all our psychological, social, economic, and ritualistic engagements are subtended by genetic codes. Economists are equally eager to propose that "market forces" are the most reliable indicators of human behavior. Historians attempt to explain events by trends, teleologies, spirits of the times, and even power relationships. The point is that each one of these disciplines takes a particular experiential phenomenal field and attempts to offer something transcendent as the ground or reason for this and other domains. But the universality of such a ground or reason is obviously restricted, since the presence of other phenomenal fields suggest the existence of other foundations. Hence, the alleged certainty of the universal claims of a given discipline or first philosophy, as an articulation of a specific range of phenomena as spatial, temporal and dynamic, reasserts the problematic of every discipline and first philosophy.[8]

At this juncture, a particular principle must be clear: As soon as philosophy and, by extension, any discipline is called on to reveal its transcending position, the depth and complexity of the phenomenal world—the world of experience—that implies such transcendence must be shown. In brief, any access to the first philosophy of any given discipline requires access to the world presupposed by those bases. That is, all disciplines are "worldly." And the same goes for art, for it can be considered a discipline intent on acquiring knowledge, as the members of the Quantum Aesthetics Group[9] and Leonard Shlain[10] argue.

This claim does not imply that we need to understand the conception of art that is about to unfold—as well as all art—as merely contingent. First philosophies, and the disciplines modeled on them, contend that the transcending explanations constitute an absolute, while the world of space, time, and movement must be contingent. Yet, it turns out that regardless of any explanation proposed by first philosophies, the world is necessarily included and hence ceases to be contingent.

However, the claim that art is "worldly" means that art needs to surrender necessarily, if grudgingly, terminologies such as "objectivity" and "subjectivity." In terms of aesthetic awareness, the reason for this surrender is the impossibility of thinking or conceiving of anything without a world (a cosmos). Yet, the latter cannot be thought to have a transcendent origin, since it is a precondition for everything that exists. Moreover, the absurdity that the world, the universe, is developing from past to present and into the future is equally nonsensical. If the universe is all that there is, then in what time and space is the universe developing? In short, the universe cannot have a cause. Hence, any causal explanation can only be applied to the objects and events inside the world; that is, any causal thinking is appropriate to only one type of awareness of space, time, and movement. More importantly, if the world as space, time, and movement is not composed of objects or subjects, or a result of some transcendent cause, then no one can speak of causes that transform how the world is understood, since only things and events can be subjected to causes. Thus, since the world is a necessary requirement for any other conception, and since art can no longer be seen as a depiction of transcendent things, then art is an articulation of the world—cosmos. This conclusion means that a transformation from one style of art to another is a leap from one way of articulating the cosmos to another. Such a finding does not mean that the subject matter of one style disappears in favor of other subject matter; rather, the subject matter is transformed to fit the new style of world awareness.[11]

In sum, the phenomenal world is an a priori for the everything else that exists. That is, nothing escapes the phenomenal, not even art. Furthermore, every event, including art, helps to construct the world. Indeed, there are different ways of rendering the cosmos, depending on how people think and portray this domain. The task, therefore, is to illustrate the different types of worlds that exist and demonstrate how the arts have facilitated their construction and transformation.

THE WORLD OF IDENTITY: RITUALS, SAYINGS, AND DRAWINGS

"Not from me but from my mother comes the tale how earth and sky were once together, but being rent asunder brought forth all things." So writes Euripides. This metaphoric sentence suggests something more fundamental: the unity of things is prior to their separation. Numerous stories, stemming from all cultures, point to an awareness of an initial identity of events, including human "artistic" creations. This is to say, for example, that each event can be identical with every other event—each saying, illustration, or dance can be the sketched through any other. Prime examples of this type of reality are the "poetic sayings" that are designed to be identical with what they say, with the very "appearance" of all things and events. What is called poetic is, at one level, associated with an "architectonic" production of the ways in which people experience a world. The poet's words establish the structure of the world and all of the events in it; poetry prescribes the ways that people live, die, love, and worship. Indeed, it establishes the places that are sacred, profane, human, and divine. Many scholars have begun to acknowledge that initially language was—and continues to be—identical with, and thus makes, the very events that are revealed linguistically.[12] This power of language is found in numerous spoken—poetic—rituals, all the way from Vedic sayings to contemporary cults. In Vedic practices, the word is experienced as capable of saving or destroying events. These outcomes are identical with poetic-ritualistic words. When a Vedic priest pronounces something, he has the power to make events happen. In fact, his very speaking is the happening. When the shaman performs a rain dance, the dance is identical with the arrival of rain. When a tribe performs a ritual of killing the sketch of an animal, this picture is identical with the traits of the animal. Furthermore, when the members of the tribe consume the animal, they acquire its powers.[13]

While the modern age is "enlightened," at the aesthetic level ritualistic practices still presume the world of identity. The poetic sayings and ritualistic practices have remained intact. Again, numerous examples can be offered. If there is a dry season, shamans—the priests, ministers, presidents, or governors—encourage the public to "pray for rain." The prayer consists of poetic sayings that, once ritualistically performed, are identical with the power of rain, or the power of the rainmaker. When one of these modern shamans says "eat of this, this is my body, drink of this, this is my blood," he or she offers a ritual that makes all followers identical with the body and power of the cult's founder. For some major personality cults, such as Christianity, not only sayings, but also the statues and paintings are not representations of entities, but are identical with these beings. People kneel before them and implore favors; the paintings and statues are carried in processions and, at times, accused of not making events happen that the population wants. In this case, the identity of an artwork, whether poetic sayings, prayers, statues or paintings, is identical with the desired events.

This identity, at this aesthetic level, extends to human individuality. Note how people claim to have an identity based on verbal designations such as "I am a president," "I am a Christian," "I am a socialist," or "I am a priest." Identity is gained from the very function, event, or entity that one enacts, speaks, and literally embodies. This type of identity can also be seen in television advertisements and sporting events. Every advertised product is surrounded by pictorial and musical imagery in order to make the product identical with that imagery. "If you buy these shoes, you will be the sports star"; "if you use this cream, you will be Cher." Nonetheless, the imagery and sound comprise the poetic ritual that makes the shoes and the overpriced cream into the power to create new persons who are identical with these products. A similar transformation takes place in sports. If some Spanish team wins a game, say in Brazil, the population in Madrid, sitting in a bar, will jump up and chant "we are number one" or "we won." The persons in the bar were sitting and watching television thousands of miles away, yet they became identical with the team. Indeed, this type of phenomenon occurs globally, including the performance of violent rituals on the streets. Fires are set, property is destroyed, and lives are lost in the ritual chant of identity: "We are number one." These aesthetic phenomena help to account for revolutionary mass movements that are led by incantatory slogans designed to make the population identical with the ideological chants and metaphorical sayings. Indeed, in such events one may become identical with the exaggerated paintings of leaders, who are the pure embodiment of ideology and revolution, as the population attempts to make itself into the image of the "leader."

Every spaciotemporal point is identical with every other point. Accordingly, measurable distances are irrelevant. A verbal incantation done "here and now" is identical with an event "there and then." The dances and prayers for rain, in every sacred space, are identical with the power of rain in every area of the sky. There is, in this identity, an "aesthetic quantum leap" that is not dependent on events occurring "one after another" or "one next to the other" in a causal sequence. This is to say, the ritual chants and dances are not "causes" that make something happen; they are identical with the happening and hence exhibit an immediate leap—without a distance—from the artwork to the event. Of course, some might argue that a sporting event in Los Angeles caused a riot in Detroit. Yet, such an explanation breaks down the laws of causality. After all, a cause must be (a) in space-time proximity and (b) commensurate with the effect, that is, the latter cannot be greater than the cause. In this sense, the artwork that is identical with the event cannot be related in any causal way to an occurrence. They are mutually identical in space, time, and movement, and leap, so to speak, one into the other.

In sum, the world of direct transposition of poetic and ritualized speaking, of incantations, dances, paintings, and statues, which allows all events to become all other events, is metonymic. Any term, event, entity, or thing can replace any other.[14] Nothing is excluded from the logic of "all with all" that promotes transposition at a distance. In contemporary terms, there is no specific "essence" of anything; everything can become everything else.

However, the world of identity is a restricted world where quantum aesthetics cannot thrive. Since everything is interconnected, this world may appear to be compatible with the quantum principle that states that "A" and "not-A" can exist at the same time.[15] However, this is only a rhetorical illusion. In the world of identity, everything can become everything else, but only because everything is everything else. That is, there are no essential differences between "A" and "not-A." Furthermore, "B," "C," and "D" are not really "B," "C," or "D," because they are really "A." In other words, in the world of identity everything participates in the *same* essence, while in the quantum world everything participates in a *common* essence.[16] In this world, everything is interconnected because all phenomena are identical with everything else. This fact renders the holographic principle—the whole is included in the part—of quantum aesthetics[17] impossible because the whole and the part are the same, and not just included in each other. If we define the quantum process of individuation as "the task of abstracting our own identity from that which is common to others; the task of discerning what belongs to us exclusively and what our community has imposed on us,"[18] then we have to recognize that this differentiation cannot happen in the world of identity. How can the individuation process happen if individual and community are the same? From what would the individual individuate?

In sum, the quantum world is a world with differences among which connections exist; this is not an undifferentiated world as is the world of identity. Consequently, the latter world is neither the one that quantum aesthetics strives to build, nor the one that may let this aesthetics thrive. The world of identity lacks the possibilities and differences inherent to quantum aesthetics.

THE WORLD OF ETERNAL RETURN

In this second type of world, unity becomes more complex and diversified. Unity is transformed into a polarity whose poles are in continuous exchange. The rhythmic style of dance is an intimation of space, time, and motion that comprises a cosmos within whose context every mode of art is articulated. Dance is a metaphor for all arts that are depicted rhythmically, in such a way that rhythmic movement composes music, painting, architecture, and sculpture. Some writers have suggested that daily affairs could be regarded in terms of rhythmic dynamics. Yet, what is also characteristic of this dynamic is that it forms a cyclical closure. In daily affairs, there is a rhythm of seasons, running from birth to life and finally to death, and from death to birth to life, thereby leading to cyclical repetitions. Thus, the entire universe moves in the cyclical repetition of rhythmic dynamics.[19] Most important to note at the outset is that in this cosmos essentialism is avoided.

For example, Buddha sits in a lotus position; his hands are poised in a rhythm; he is neither going up nor is he pulled downward. He is centered on himself, but not as a point; rather, he is a center for the emanation of energies by whose force the entire universe dances. He is expressive in the sense of spreading benevolence, tranquility, and an all pervasive understanding that all events, including the human, are in a flux of coming and passing, appearing and disappearing, and coming back again.[20] The cosmic rhythm makes human life light and easy, playful, and accepting of all the vicissitudes and tensions that come and vanish. Indeed, all Indian depictions of eminent beings are premised on the cosmic dance, musicality, the breath, the beat of the drums of creation, and the eternal return of all events. Thus, Shiva is a figure that depicts the dance of the universe, while Brahma, the all-pervasive one, is literally the rhythm of breath and soul, inspiration and expiration. In this rhythmic-cyclical world, equal to the preeminence of dance is sound, whether chant, song, cadence, or the all-pervasive cosmic "aum" that moves from sound to silence and returns to sound. However, no one should assume that the physical composition of sound, and its extension to human speaking as oral, is the basis of the awareness of the cosmos as rhythmic and cyclical. Sound and silence, just as much as dance or the rhythm of waves, the cycles of the planets, the sun, and the seasons, are equally employed in aesthetic creations to articulate this type of world awareness.[21]

Associated with the cosmos of rhythmic cycles—the eternal return—are expressive aesthetic dimensions that constitute "mood space." Such a space is based on the notion that rhythms, and their audial variants, carry with them moods, such as excitement, violence, or eroticism, in a way that the entire environment is understood to be pervaded by these moods. Indeed, the latter draw everything into their imageries. Most important to note is that these expressivities are not yet associated with objective or subjective meanings, but with participatory dynamics. Hence, a particular musical rhythm moves a person to join the dance, the chant, the humming of a tune, or the excitement of the environment. In brief, in this cosmos and its space and time dynamics, there are no "neutral" things.

There are also arguments suggesting that this cosmos is "psychologically laden"; yet, the psyche is not something projected by persons. The reason for this prehuman conception of the psychological lies in the fact that human bodily participation in the rhythmic and cyclical cosmos is not as extensive as the participation of these rhythms in human bodies. In this sense, humans get caught up in a specific living of time, space, and movement, and depict these dimensions in their aesthetic expressions. Accordingly, art, in this cosmos, describes major figures in terms of characteristics or dimensions that are all pervasive. Hermes is the solitary and solace of the night; Artemis is the tenderness and softness of the environment; and Aphrodite is the wildness of the world.[22] The same can be said of the stories from the East, where the great figures, such as Shiva, are depicted in terms of their cosmic expressivity as Kama—Eros or Lila—plays, or the sages who sing of the great heroes and thus envelop them in a space of moods in which the public participates. While this dimensional space, time, and tensile movements are dynamic and polarly shifting, they have no orientation, and neither do those who participate in these changes. This can be seen in dance and music. Dance does not move in a linear fashion; the rhythms, so to speak, have no teleological direction, and hence do not aim at anything, just as sound and music, with their all-pervasive and overlapping volumes, are not "going anywhere."

Paintings and, as already suggested, plastic arts, are also used to depict this cosmos. In medieval paintings, there is a striking lack of modern spatial and temporal characteristics. These paintings possess a depth that is expressive, a frame that is vaulted—the rhythmic cyclical aspect—and a central figure from which mood dimensions radiate. All figures, surrounding the central figure, participate in the mood that is also designed to draw the viewer into its sway. What is important are the ways in which the moods are captured by the polarizing movement of imageries. For example, angelic figures float down, and "sinners" strive to move upward in an eternal drama to shift polar positions. While there are central figures, such as the crucified or the Madonna, they, too, are stretched between various moods: adoration, suffering, subjection, elevation, sorrow, and joy. This rhythmic and cyclical world is also obvious in the architectural designs of sacred buildings with their vaults that establish a cosmic mood of "spirituality" and tranquility, and at the same time reveal threats from demonic figures: damnation and salvation, or torture and joy are deployed in polar manner.

The eminent literatures and stories in this world depict events—including human actions—as a bifurcation, an initial separation, that offers the separated beings an exchange of positions and functions. What is high and divine becomes low and demonic, while the demonic becomes divine. What is light becomes dark, and the dark becomes light, and in such a way that one cannot exist without the other. In fact, one contains the other. For example, movement toward love is also a movement toward hate; a movement toward sound takes silence as a requisite polar aspect. The same type of dynamics can be found in the grand myths of the "fall of man." These types of myths, which abound across the globe, narrate how at first there was a paradisiacal unity—identity—of the human with the source of the world, and how the human transgressed this identity and had to decline in stature. The human goes from good and beautiful to evil and ugly. Yet, through hard labor the human will rise again from evil and ugly to good and beautiful.

But what must be avoided is a dualistic interpretation of this movement. The dynamic polarity that dominates all arts of this world cannot be rent asunder. That is, it needs to be understood as a whole, just as male-female juxtaposition, each containing the other as a polar counterpart, cannot be understood one without the other. In this world of eternal return, a movement exists that no longer identifies point for point anything with anything, but recognizes polarities that need one another in order to be viable. While one event or concept is transformed into its polar aspect, one is not given without tracing the other aspect. This means that "identity exists in difference," such that the identity becomes also a trace of alterity, just as hate is a trace of love and the demonic contains a trace of the divine.

In any case, this leap from identity to polarity does not invalidate the former type of consciousness. The logic of identity, of "everything with everything," of every word with an event, of every ritual with the power of the thing performed, is not gone. This understanding becomes a background phenomenon that at times reappears as significant. For example, in the polar movement of the "fall of man" there is also a background assumption wherein man will become one with the paradise lost, or will become truly human after the historical process moves persons from the fallen state, for example, from primitive communism to the utopian salvation where the human will become identical with the community.

In sum, in the world of rhythmic cycles there are no fixed, reified things and their corresponding "neutral-objective" empirical characteristics. All things are musical, enchanted, dancing, returning, and vanishing, and hence in constant movement from polarity to counterpolarity. One could call such a recurrent shift "periodicity."

If the world of identity was too constraining for quantum aesthetics, the world of eternal return is more compatible with the quantum ideas, even though this viewpoint still lacks the breadth that these conceptions need. The antidualism inherent to this world of eternal return is what makes it compatible with the quantum principle of indeterminacy.[23] That is, the fact that good and evil, light and darkness, summer and winter, and love and hate exchange positions and essences means that things and events lack a fixed identity—they are connected to other things and events with which they exchange essences. Thus, in the world of eternal return, identity exists in difference. However, the differences that this world can promote are limited. In other words, "A" can only be "not-A," but never "B" or "C." An exchange among the essences of the various dichotomies does not take place: things and events do not form a common whole. In sum, the quantum principle of nonseparability[24] does not apply to a world that is fragmented into a myriad of dichotomies. Thus, even though there is certain compatibility, the world of the eternal return lacks the depth that quantum aesthetics prescribes.

THE WORLD OF PERSPECTIVE AND SEQUENCE

The sequential and perspectival universe is premised on the idea that both space and time have recognizable points on a line that can be identified, fixed, and accessed from the positional point of an observer. This positioning requires that anything without a position, anything that is adimensional, such as moods and expressivities, must be reduced to characteristics of an object. Expressive dimensions are relocated into a "subject" who is also an ego and who construes the world through a fixed perspective. This means that the permanent ego, separate from other egos, requires the establishment of a sequentially and perspectively positional cosmos, such that things move from past to present, and from here to there. The ego is recognized as having a spaciotemporal position, from which other spaciotemporal points can be accessed. Other positions, occupied by other egos, allow them to be equally separate individuals. Again, the ego becomes the topos where the expressive world of mood dimensions become located as "inner" states. This new cosmos requires the transformation of the dimensional, expressive world into a subjective psychological state that, accordingly, can be projected onto impartially located things. Here is born the essentialism that allows things to have their determinable—eternal characteristics and causes—spaciotemporal locations and sequences. In this cosmos, individuals are not transferable, since they have their identifiable space and time positions that no one else can enter. The identity of each individual is premised on the identities of fixed locations. What happened in the past, for example, can be recognized as a designation of a specific point and spatial location.[25]

The quantum leap from the polar cyclical cosmos to the perspectival sequential universe is aesthetic to the extent that artists and the poets—prior to scientists—have sketched this universe. That is, the perspectival space that gave rise to positional views on positioned objects, and positional points on a time line, was first painted, sculpted, and depicted by artists. Later on, sciences adopted this universe a priori, as if it were too obvious to need further explanation. In this sense, the objectification of space, time, and movement is not value free, but is based on a specific view of aesthetics. The scientific or atomistic individualism that could be fragmented into detailed analyses could not be obtained without the background of this a priori universe. Moreover, this universe also allows the continuous fragmentation of space, time and movement, thus leading to the incrementation on spaciotemporal perspectivity. This fragmentation is obvious in the increase in disciplines. Each discipline has its own perspective that can be split into subperspectives, where each "sub" can become an independent perspective. At the outset, this universe calls for reification. All things must be "material" and thus divisible into parts, perhaps into infinity.

This type of cosmos is the basis for the famous modern dualisms. Perhaps the best known is mind-body dualism, followed by the subject-object, the inner-outer, and the fixed hierarchical high-low, horizontal left-right, and past-future divisions. Yet, the logic that pervades these dualisms is "either/or." Something is past or future, mind or body, left or right, subject or object, and light or dark. These strict separations exclude any polar shifts. In dualism, something is either male or female, and if a person does not belong to only one of these categories, say an androgynous human, then such a person is "abnormal," an anomaly.

The language of this dualism, in contrast to musical, dancing, and rhythmic, is premised on light metaphors. Indeed, the artists who traced this cosmos were, almost without exception, completely enamoured with light and shadow. All events, including the reified and positioned things, have to be understood in the "light of reason," appear in optimal clarity, be articulated by clear and distinct rational laws, and be seen from an indifferent vantage point. Metaphors related to light pervade all arts. In the plastic arts, for example, space is demarcated by perspectives drawn from a center, similar to the way in which a precise time point is depicted by a shadow cast by a thing. Music, for its part, is no longer rhythmic but a sequence of hierarchically arranged notes. Additionally, writing depicts a sequence of events and not a dynamic transposition of images, while all the wild rhythms and chants are relegated to the nether region of darkness, basically as psychological and inner torments of souls that require clear illumination by scientific reason.

Another important dualism is premised on the separation of space and time: space is articulated in terms of the deployment of material components, while time is the measure of this deployment in sequence. Thus, in principle, this cosmos is dualistic because the two fundamental aspects of the general cosmic a priori, space and time, are separated. This separation leads to a static conception of all things, specifically when temporal movement is a measure separated from spatial events or objects.

In the perspectival world, space and time follow fixed lines that can be divided into fixed points where all elements are located. Everything is severed from everything else and acquires a unique and fixed identity different from other identities, for every element is located in an unmatched position in both time and space. The world of perspective is such that "A" and "not-A" cannot exist at the same time. Contrary to what quantum artists and critics would claim,[26] "A" and "not-A" are two different essences that are located in different places within space and time. Consequently, individuation cannot happen in this world: From what can we individuate if everything is different and nothing is common? The perspectival is a world of individualism, but not of individuation.

This is the world of dualism where everything is fixed, possibilities are very limited, and mind and body, matter and energy become distinct and separate. In this world, *mental matter*—the term quantum critics use to refer to a reality embedded in human subjectivity[27]—cannot exist. In other words, the perspectival world is contradictory to the "anthropic principle" of quantum aesthetics.[28] The human does not position the universe. Quite on the contrary, the universe has its own autonomous coordinate system within which the human is positioned. In this world, the human being is only the *origin*—with a little *o*—of perspective, but never its *Origin*—with a capital *O*—or first and immanent cause, as supporters of quantum aesthetics claim.

In sum, the perspectival world is a realist domain in which mind and body are differentiated and become distinct. This world is a Newtonian and Balzanian world whose corpse Gregorio Morales wants to bury, because it is restricted, naive, antidemocratic, and lack possibilities for individuation and socially responsible creation.[29] This is thus the type of world against which quantum aesthetics, with its belief in the integration of opposites, is reacting.[30]

THE WORLD OF DYNAMIC FIELDS

Artists and philosophers broke away from the dualism present in the world of perspective by merging time and space, and thus conceiving of time not as a measure of sequences, but as a dynamic field of relationships. This development suggests that every position is a constant "positioning" in a field of other positions, such that with a shift of one, others are shifted and transformed into a field of phenomena. A writer, for example, cannot pluck out something from his or her surroundings and offer a perspective on it, without changing this factor and the surrounding field.[31] How can the size of an element be understood without also asking about the context that allows this particular thing to have a specific size? All events have their copresence in a field prior to any differentiations that are made with regard to spatial and temporal locations. Moreover, events contextualize themselves in such a way that they "overlap" in many ways. A rose is bright red due to the hazy yellow tulip next to it, and this hazy yellow is possible due to the intensely green bush that is also present. Each quality shades across the others, not as if one thing caused another in a sequence but as shifts of the entire field. In this cosmos of space-time dynamization, there are no discernible focal points, such as the center of a perspective or a shadow designed to locate a point in time. Thus, persons should no longer speak of a past that is gone and a future that is coming, or of something being here and something else being there. All these features are field phenomena, and thus to speak of a future is to find it in a field that contains the past and present, or the here and there. This finding suggests that all events comprise a field of correlated differentials. In other words, every event is not a polar aspect of another event, but is transparent with the surrounding differences. The term "transparent" indicates the copresence of events, such that one can be seen only through the other as different. Thus, something is different only with regard to the copresence of the others; there is no total identity of any of them, and yet each one is identifiable through the copresence of the differences revealed by others. If one shifts, the others shift.[32]

The world awareness as a field in art appears in the temporal dynamization of space, particularly the disruption of perspectival positionality. Famous buildings lose their shape and burst through clouds, dancers disappear in a whirl of loud and subdued colors, vortexes of energies form and deform intense shapes, a wall painting "breaks down" the Euclidean geometry of space, and figures deployed on a meadow are traces of curved space. The well-respected metaphors of things as Euclidean and positional are abandoned in favor of the language of flux. This gambit, however, does not entail a total dissolution of recognizable phenomena.

The world of dynamic fields consists not only of the dynamization of space, but also the disruption of perspective by overlapping phenomena. Qualities are detached from things, thereby giving things an ambiguous and diffused appearance. Such diffusion dislocates positions and perspectives and creates field depths that do not seem to have fixed boundaries. Indeed, the very shape of things shifts with the shifting of overlapping qualities. One reason for detaching qualities from things, and even events, is to demonstrate that the very shapes of things are possible in a dynamic spaciotemporal media whose shifts also alter the appearance and shapes of objects. The shapes of things emerge from and drift into these dynamic media and thus lose their stability, like floating lilies. Surfaces lose their "thing like" characteristics and yield to depths of phenomena seen one through the other, but constantly destabilized.[33]

Furthermore, a spaciotemporal field can be depicted by showing something from "all sides" at one glimpse. Remember that sequential perspectivity presents things from one perspective. In order to grasp another perspective, the viewer has to constitute a series of other perspectives, from different sides of the "same object," and thus engage in a sequential movement. After all, in that cosmos, one cannot see a thing from all sides at once. Yet, artists broke down the sequentiality of perspective and were able to dynamize figures, so that all sides are transparent and each is revealed through the others. The logic is that any perspective exists in a field that must be seen "all at once" through the copresence of other, different perspectives and times. This is exactly what cubist painters do.[34]

Such differential phenomena also appear in music. Recall that in the world of sequential perspectivity music consists of a series of fixed notes and patterns of sounds. However, for the artists who understand the cosmos as field awareness there are no perspectival melodies, but atonal explorations that seem to disrupt the usual "order" of sounds. Indeed, even the polar rhythms are abolished in favor of arrhythmic compositions—diaphonies—that never return to the same "spot" or combination. This opening of musical sounds includes their synaesthetic character or the pure depiction of audial colors without the articulation of expressivities or realities. This is to say that the medium consists of pure space-time dynamics that have no aim, teleology, or resolutions. The point is that this musical medium explores audial field differentials in every possible variation, without prejudgments about form and content. Another description might be that the atonal and arrhythmic music is a concretization of time as a dynamization of space. One notable characteristic of this music is that it is not designed to "inspire" listeners. In the cosmos of rhythmic cycles, music was connected to "muses" who provided inspiration to persons. In the cosmos of field differentials there are no muses, and there is no way to join the rhythm either by tapping of the feet, chanting, or dancing. How would one dance to Anton Weber's music?

The dissolution of fixed positions and substantive characteristics is also apparent in poetry and literary texts. There are new theories of language, entitled "field linguistics," where neither terms nor statements are the basic units of language. Every so-called term or statement can be understood only within the entire, open, and shifting linguistic field.[35] Writers and poets extended the field notion of language by exploding the substantives. Nouns have always played the role of designating fixed realities that have permanent characteristics, while verbs were designed to indicate the way in which these substantives act. Now, the substantives are verbalized, become active, and lose their permanent features without requiring verbs. This is to say that the very notion of "subject-predicate" becomes redundant. If this schema is still used, then there is a full recognition that its use is a matter of convenience and not an exact delimitation of reality. Things now are neither absolute nor relative, neither finite nor infinite; events are depicted as transparent or differentials with depths and horizons that are "thick and ambiguous." No longer should critics look for propositions that have only one meaning, but, on the other hand, they should not announce that meaning is impossible. And the same can be said about the subject, who is not dead but transparent.

The art based on the field cosmos decenters the subject from its supreme position as the source of value, beauty, and meaning, yet this decentering does not abolish the subject. A viewer of a particular scene may seem to take a position and a perspective, nonetheless the events surrounding this scene are also "taking positions" with respect to the scene and the viewer. The person is also seen by the positions of events. A viewer may assume a position with regard to this side of a building, yet the other side is already revealed by the tree behind the building, the shack beside the building, and the cloud above the building, such that this individual is positioned as one aspect in a field of differentially copresent visions. As persons locate events, the events locate them. In this sense, the subject has been decentered from its absoluteness, but has not died.

An individual is empowered to decenter other events, while they are empowered to decenter the subject. This copresence of events should not be described as a cause-and-effect sequence, since no component can be understood as coming first or second. The subject, in its movement around a building, is repositioning other events in the field. But the subject is not a cause of this repositioning, since these events also reposition the subject. Hence, there is no cause-and-effect dualism in the field cosmos.

In sum, the world of dynamic fields is profoundly antidualist. Everything becomes transparent and revealed in this montage of interconnected identities. Thus, all events are always in progress, for a minor change in one of them provokes a change in the entire field. Regardless, neither subjects nor events disappear; they each presuppose the existence of the other.

The world of dynamic fields is undoubtedly the world of quantum aesthetics. This is a world full of possibilities in which, according to the hologramatic principle,[36] everything can be seen in everything else. Indeed, everything is transparent and reflects everything else necessarily. The essence of things and events is always a negotiation among the various objects and individuals that interact with them. In quantum terminology, this is a world in which "A," "not-A," and even "B" interpenetrate one another.[37] This world is also a domain in which, according to the anthropic principle, human beings have a decisive role in the definition of "A" and "B."[38] Contrary to what happened in the perspectival world, in the world of dynamic fields things and events are not located within an autonomous set of spaciotemporal coordinates. Rather, they are located in the world by humans who are, in turn, located by other humans, things, and events in a dynamic process. Thus, compatible with quantum aesthetics, in the world of dynamic fields things do not have fixed identities.

In sum, causality is not possible within dynamic fields, for time does not follow a straight line. Consequently, things cannot be said to be caused by other things, rather they are said to have a certain synchronicity[39] with one another. In other words, the world of dynamic fields is the world that quantum aesthetics is trying to build, while fighting against the realism and dualism that is characteristic of the perspectival world.

CONCLUSION

There are fierce battles among artists and art critics concerning the "grounds," sources, causes, and reasons for art. In this brief chapter, every explanation of art could not be addressed in detail. Rather, the focus of this discussion is the worldliness of all artworks. By this term is meant that under any reading of art, the world, or cosmos, is given a priori. Without the world as an a priori, it is impossible to think of things, events, their multifarious relationships, and their appearance and disappearance. Following this, the main thesis is that while things can be made, built, invented, or created, the cosmos cannot. There is no cause for this cosmos, but only for things and events within one type of cosmos. Art has helped create various types of worlds—the world of identity, eternal return, perspective, and dynamic fields—where events and things are understood in different ways. Indeed, art is responsible, like science and other disciplines, for the creation and maintenance of these worlds.

Furthermore, the change from one artistic style to another very often signals a leap from one world to another. Quantum aesthetics can be said to be both somewhat compatible with the world of polar return and a driving force in moving away from the world of spaciotemporal perspective into the world of dynamic fields. If the world of perspectives is dualistic and rigid, the world of dynamic fields is antidualist and flexible and the perfect environment for quantum aesthetics to thrive. Furthermore, this world of fields is the one that quantum artists labor to create.

NOTES

1. Algis Mickunas, "Moritz Geiger and Aesthetics," in *Analecta Husserliana*, vol. 26, ed. E. F. Kaelin and C. O. Schrag (Dordrecht: Kluwer Academic, 1989), 46.
2. Thomas Langan, *The Meaning of Heidegger* (New York: Columbia University Press, 1961), 176.
3. Jean Gebser, *The Ever-Present Origin* (Athens: Ohio University Press, 1984).
4. David Bohm, *Wholeness and the Implicate Order* (London: Routledge and Keagan Paul, 1980).
5. Ibid., 124.
6. Martin Heidegger, *History of the Concept of Time* (Bloomington: Indiana University Press, 1985), 171.
7. Immanuel Kant, *The Critique of Pure Reason* (London: Encyclopaedia Britanica, 1952), 29.
8. Fred Kersten, "Multiple Realities in Literature," in Lester Embree, ed. *Alfred Schutz's "Sociological Aspect of Literature": Construction and Complementary Essays* (Dordrecht: Kluwer Academic, 1998), 152.
9. Quantum Aesthetics Group, *Quantum Aesthetics Group's E-m@ilfesto*, 22 July 2000 ‹http://teleline.terra.es/personal/lucschok/estetica/emailfestoeng.htm›.
10. Leonard Shlain, *Arts and Physics: Parallel Visions in Space, Time, and Light* (New York: Morrow, 1991), 15–16.
11. Jean-François Lyotard, *Postmodern Fables* (Mineapolis: University of Minnesota Press, 1997), 24.
12. Walter Rehm, *Orpheus* (Darmstadt: Wissenschaftliche Buchgesellschaft, 1972), 75.
13. Mircea Eliade, *Patterns in Comparative Religion* (New York: New American Library, 1974), 216.
14. A. Julien Greimas, *Du Sens ii* (Paris: Editions du Seuil, 1983), 35.
15. Quantum Aesthetics Group, *E-m@ilfesto*.
16. Ibid.
17. Ibid.
18. Gregorio Morales, *El Cadáver de Balzac* (Alicante: Epígono, 1998), 25. Translated by the editors.
19. Sarah Kofman, *Nietzsche and Metaphor* (Stanford, CA: Stanford University Press, 1993), 6.

20. Dagobert Frey, *Grundlegung Zu Einer Vergleichenden Kunst-Wissenschaft* (Darmstadt: Wissenschaftliche Buchgesellschaft, 1970), 25.

21. Alan Watts, *The Two Hands of the God: The Myths of Polarity*, vol. 3 (New York: G. Braziller, 1963), 82.

22. Ludwig Klages, *Grundlegung der Wissenschaft Vom Ausdruck* (Bonn: H. Bouvier u Co. Verlag, 1964), 114.

23. Quantum Aesthetics Group, *E-m@ilfesto*.

24. Ibid.

25. Rollo May, *The Cry for Myth* (New York: Norton, 1991), 108.

26. Quantum Aesthetics Group, *E-m@ilfesto*.

27. Morales, *El Cadáver de Balzac*, 22.

28. Quantum Aesthetics Group, *E-m@ilfesto*. See also John Barrow and Frank Tipler, *The Anthropic Cosmological Principle* (Oxford: Oxford University Press, 1988), Chapter 10.

29. Morales, *El Cadáver de Balzac*.

30. Quantum Aesthetics Group, *E-m@ilfesto*.

31. Karl-Otto Apel, *Toward a Transformation of Philosophy* (London: Routledge and Kegan Paul, 1980), 136.

32. Maurice Merleau-Ponty, *Signs* (Evanston: Northwestern University Press, 1964), 170.

33. Morris Berman, *The Reenchantment of the World* (Ithaca, NY: Cornell University Press, 1981), 153.

34. Gebser, *Ever-Present Origin*, 476–479.

35. Lothar Schmidt, *Wortfeldforschung* (Darmstadt: Wissenschaftliche Buchgesellschaft, 1973, viii.

36. Quantum Aesthetics Group, *E-m@ilfesto*.

37. Ibid.

38. Ibid.

39. Morales, *El Cadáver de Balzac*, 124.

CHAPTER 5

Quantum Aesthetics and Art History

Jennifer Wilson

"[T]oday, as a result of a better survey of Time and Space . . . there may be the same basic and universal force operating which, since the dawn of the world, has constantly striven towards an ever-growing organization of Matter. We must no longer think of this force as a mere spatial motion of the Earth (Galileo), but as the tightening, beyond ourselves and above our heads, of a sort of cosmic vortex, which, after generating each one of us individually, pushes further, through the building of collective units, on its steady course towards a continuous and simultaneous increase of complexity and consciousness."

—Pierre Teilhard de Chardin[1]

STORIES AND MEANING

We feel as though we will burst sometimes with the sense that our lives are filled with a deep, rich meaning that we cannot easily articulate. We can't explain the welling intuition that tells us there is an undeniable connection between ourselves and some universal design. Our hunch that we are somehow woven into a grand tapestry of meaning is simply ineffable. So, we pass on stories we have heard of amazing "coincidences" and uncanny occurrences in the hope that these tales will convey what we ourselves cannot.

> A young girl bravely approaches the high fence of a concentration camp in her small village in Poland. She throws an apple over the barbed wire to a young boy inside the camp. He eats it hungrily and returns to the same

spot the next day to receive another apple. Each day, the young girl on the other side risks her life to toss the young boy an apple. One day, the boy is shipped off to another concentration camp. He has just enough time to tell the young girl that he has to go and that she has probably saved his life by bringing him food. He tells her he'll remember her kindness forever. About thirty years later, in New York City, a man is set up on a blind date. As soon as his date gets into the car, the man recognizes her but does not know from where. His entire body is ablaze with a sense of recognition and deep love for this woman he's supposedly meeting for the first time. Finally, at the end of the date, the woman tells the story of throwing apples to a young boy in a concentration camp in Poland. The man bursts into tears and tells her he's that same boy and in a rush of emotion proposes to the woman. They get married and are still blissfully together.

Stories like this deepen our belief in life's rich significance, and by passing them on we strip away the veils of the universe's great mysteries. We hear stories like this and are compelled to share them with people close to us. Similarly, we pass on stories of our own coincidences, of our very own prescient dreams and chance encounters. We listen intently to stories of survival and miraculous healings. A young woman's brain stem was damaged in a car accident; her doctors concluded she would survive only on life support and for a short time. Her mother hung a sign in the Intensive Care Unit that said: "No negative things to be said here." Her daughter recovered fully. Think of your act of listening right now, to that story, and how you feel. Stories of miracles and survival ignite in us the latent, deep conviction that there is a meaning, a shiny, ineffable truth about the universe to which we are connected; this belief feels as simple and beautiful as a ring, and yet is so difficult to articulate, to explain. Stories of hidden, rich meaning draw us closer to the core of what must be a universal design.

ART

Beauty beckons us in the same way. Shape, color, line, curve, arch, dent, angle, light, and space all draw out of us some kind of soul reaction, some kind of visceral response that is beyond language and mind and transforms everything we know. A woman stands in front of a Mark Rothko painting. Physically speaking, only her retinas are basking in the fields of humming, floating color; yet, in that moment, there is a deep conversation, a stirring, between something within her and the painting, a connection that she cannot understand intellectually, but can feel deeply, unquestionably.[2] And at the moment that she experiences this stirring, her capacity for understanding (in a larger, beyond-the-mind way) grows. The stirring, soul-reaction to the painting has expanded her, her consciousness, her awareness. The mind is a contraction of our larger capacity to grasp universal realities. Art, like inspiring stories, pulls us beyond the smaller mind to our natural capacity for more expansive awareness.

In a recent interview with Wendy Lesser, choreographer Mark Morris touched on this very idea. Lesser asked Morris if he had been influenced by Merce Cunningham, perhaps one of the most influential choreographers of all time.[3] Morris laughed and told an amusing story. He said that he used to go to see Cunningham's dances for the same reason that people used to take fish oil pills in the 1980s. People took them because they had heard that it was good for them, even if they did not quite understand why. In the same way, Morris went to see Cunningham's dances on a regular basis; they were good for him, for his growth as a choreographer, although he did not know why. Then, during one performance, Morris said he suddenly and completely "got it," and then said "you know, I mean I 'got' it in the way that you 'get' the ocean." That is, "getting it" is not a function of the *mind*, logic, or deduction; instead, something else happens, there is an expansion of awareness, a deep wash of understanding that doesn't feel like "thinking" yet can alter how you think and "understand" something forever. This moment of expansion, of a deeper understanding, is what art has to offer us.

We suspect that there is some beautiful, simple order to the world and that we are somehow a part of it. Like the stories we pass along to keep that belief alive, certain works of art hold us in a long, still moment when we can feel the contours of our souls; we can experience the meaning we are all so drawn to and yet cannot explain, and in these long, still moments we can begin to "get" it.

PHYSICS

Consider the fact that Claude Monet spent the majority of his life devoted to chasing light. He was obsessed with the beautifully indefinable nature of light and what its fleeting characteristics revealed about the universe, time, space, and experience. Light, for Monet, was a doorway that he walked through his entire life, and his experiences with the nature of light were the ongoing defining moment for his entire artistic career.[4]

At the same time, Albert Einstein pursued the nature of light with fervor, tracking its countless characteristics and properties while he tried again and again to unleash it from the tethers of classical science.[5] Einstein was possessed by what he saw in light, and while Monet continually sought the best way to paint it, Einstein devised new, revolutionary ways to calculate it. Coincidentally, Einstein and Monet were both contemporaries and conceptual revolutionaries of the early twentieth century. During that period in the history of art and science, the concept of light changed forever and they altered the way people would think about, dream about, or "get" art and physics.

ART, PHYSICS, AND CONSCIOUSNESS

Thinkers such as philosopher Pierre Teilhard de Chardin have argued convincingly that consciousness is expanding on a universal scale.[6] He believed that there exists a universal consciousness that is on an evolutionary path. He also believed that art plays an important role in this progressive expansion as it increases our capacity to understand universal truths.

Scientific discoveries of the past one hundred years have reached similar conclusions. From relativity theories to quantum physics and string theory, scientists are closing in on (or opening up to) one equation to express the nature of the universe. New scientific discoveries have given us a fresh way to perceive the world and our connection to the universe as being beyond duality and determinism, beyond the confines of logic, linearity, and materialism. In the same way, the new sciences reveal the deep connection between art, science, and the pursuit of the universe's mysteries. Consider the striking resemblances in the following: Einstein's quest to define the nature of light, Stephen Hawking's relentless search for a unified field theory, Monet's lifelong pursuit of light's essence, Paul Cézanne's private obsession with Mont Sainte Victoire, and the soul's yearning for the ocean. They all speak to the same essential longing—to understand our connection to the universe and to grasp what the basic elements (such as light) can reveal to us about that connection. The pursuits of science and art are both expressions of this innate longing.

Consciousness is on an evolutionary path. Art history and science are on evolutionary paths. Art accelerates universal consciousness along its evolutionary path. These are some of the precepts of the quantum aesthetics.[7] If we embrace the notion that collectively we are on a path of consciousness evolution and that art is an expression of as well as a catalyst for that evolution, then we can begin to look at art and the history of art from a new perspective. Art is the expression of a *zeitgeist*, the spirit of the time, and is characterized by a state of collective consciousness.

In his book *Art and Physics*, Leonard Shlain draws some staggering connections between revolutionary events in art history and significant scientific discoveries.[8] His premise is that artists have always taught the world to "see" what scientists teach the world to "understand" about universal truths. In a fascinating trek through the history of Western art, Shlain shows how artists have presaged scientific discoveries. Time and again, artists were ahead of scientists in their discoveries and interpretations of concepts such as gravity, the relationship of time to space, and human physical experience. The before-mentioned similarity of Monet's and Einstein's pursuits of light is one example of this trend. Now consider the compelling connection Shlain reveals between Isaac Newton's and Leonardo da Vinci's shared obsession with motion. Leonardo and Newton came to miraculously similar conclusions about motion based on their own studies and experiments.

According to Leonardo: "All movement tends to maintenance or rather all moved bodies continue to move as long as the impression of the force of their motors (original impetus) remains in them."[9] According to Newton, "Every body continues in its state of rest, or of uniform motion in a straight line, unless it is compelled to change that state by forces impressed upon it."[10]

Leonardo, one hundred years before Newton, presaged the scientist's discoveries about the nature of motion and our experience of it. What Shlain points to in his revelatory connections between scientific discovery and art history is similar to the basic premises of the quantum aesthetics: that consciousness expands on a universal scale, that science and art (especially the impact of aesthetic properties on individual experience) can be understood in similar terms, and finally that the link between the expansion of universal consciousness, the affect that art has on this expansion, and the unfolding of a scientific understanding of the universe, is profound.

ROADMAP

In what follows, I will elaborate on the basic premises of quantum aesthetics to show that not only have artists' works anticipated significant scientific discoveries, but that they also have helped us to perceive the world in a new, more meaningful and expanded way and thus have affected and continue to affect the expansion of our collective consciousness. I will show how impressionism and cubism sealed the fate of seventeenth-century worldviews and offered a new way of thinking about art, science, and man's relation to the world that transcends the more limited concepts of dualism, determinism, and causality. Quantum art suggests a new reality that is neither causal nor reductionist. I will try to show that, given this, the quantum aesthetics is a general philosophy with social and political importance. Finally, I will demonstrate the place of quantum aesthetics in the history of Western art.

DOUBT, DUALISM, AND ROMANTICISM

In some ways, we are still in the throes of the romantic era and are trying to reestablish the connection of mind/body/universe after their harsh separation that was the direct result of seventeenth-century thinking. What has become known as the seventeenth-century worldview can be described by the three main modes of thought that developed out of this era and set the stage for what resulted in modern angst, namely: dualism, determinism, and mechanism.

Mind-body dualism is the belief that the mind and body are completely separate entities to the extent that they consist of totally different properties.[11] René Descartes, the philosopher famous for the quotation, "I think, therefore I am," is considered the father of this mode of dualist thinking. For Descartes, the inner realm (imagination) is different and separate from the outer realm (nature). The mind is capable of rational thought, feeling, and imagining, while the physical world is reducible only to the strict laws of science (mechanism). Descartes and his colleagues of the scientific revolution of the seventeenth century brought about a view of nature and the universe that could best be described as an inanimate, uninspired machine.

Isaac Newton, who mastered Descarte's theories and became the leading mathematician in Europe before receiving his bachelor's degree in 1665, upheld these sharp distinctions, as well; but he also believed that truth could be found in nature. Newton's laws of nature, however, were also mechanistic because they were reducible ultimately to the laws of mathematics. The seventeenth-century worldview, again, best described by the precepts of mechanistic thinking, were etched and elaborated on by Descartes and Newton.

The belief that the physical world, including (to some) the human body, could be seen and understood as similar as a machine, along with the belief that the human mind was at best a tool for rational and linear reasoning (not to be distracted by the imagination), set the philosophical and scientific stage for the industrial revolution that followed a century later. If history unfolds like a narrative, then we could say the mechanistic and dualist worldview of the seventeenth century was a prologue to the eighteenth century's dramatic climax—the industrial revolution's formation of cities, factories, mass production of goods, steel machines, the dismantling of farm subsistence and community, and the beginning of what we would later see develop into modern alienation.

William Blake posed the most radical opposition to Newtonian and Cartesian thought, and, more importantly, the industrial revolution. Artist, visionary, and poet, Blake believed that the imagination was the residence of "God."[12] Moreover, his art showed the presence of God revealed through the power of the imagination, specifically the imagination's ability to hold opposites, such as heaven and hell, or "A" and "not-A." As a proliferate writer, poet, and painter, Blake was the visionary who set the romantic era in motion. His notion that you could see the "world in a grain of sand" and "heaven in a wild flower" was a radical refutation of dualism, determinism, and Cartesian doubt.

Blake is still seen as one of the greatest artistic and philosophical visionaries, and his *Marriage of Heaven and Hell* is still one of the greatest challenges to the belief systems of the seventeenth century.[13] He interpreted the role of art in the expansion of consciousness in much the same way as Teilhard de Chardin, not as simply an expression of form and space, but as an aesthetic medium or catalyst of meditation, a means by which to experience a higher state of consciousness and ultimately enlightenment.

When we consider that Blake's work directly rejected the dominant scientific and philosophical beliefs of his time, and that the physical realities of cities, factories and machines were springing up all around him as physical proof of the ideas he refuted, we can see the real extent to which he was a visionary and how far ahead of his time his early romantic notions actually were. Critics have argued that Blake's work is still ahead of *our* time. Even the precepts we struggle to formulate in this book, to articulate a theory called the "quantum aesthetics," are clumsy and bound by unwieldy syntax compared to the lucid concepts of unification that Blake's art and poetry set free hundreds of years ago. We are only now, as a result of embracing Eastern thought and unified field theories, approaching the precepts that he so beautifully set forth. Certain movements along the way have precipitated our turning away from classical physics and adopting more open-ended notions of physical reality and the boundless potential of the human imagination and consciousness.

IMPRESSIONISM AND THE END OF DUALISM

Consider the duality of light principle: The "true" nature of light can not be measured.[14] Sometimes light appears as a wave; sometimes it appears as a particle. Light always "is" both things. But when we attempt to measure light, the framework imposed by the very act of measurement has the effect of squeezing this elusive element into the shape of only one of its forms. In the case of light, the scientific method only gives us one small sliver of its rich story. According to this principle, the *wave* and the *particle* are two different descriptions of physical being that are equally valid at the same time. "A" and "not-A" exist at the same time. Light is simultaneously "A" and "not-A."

The more comfortable we get with the multiform properties of the most basic physical makeup of the world, the sooner we can begin to understand the true nature(s) of existence. Impressionist painting is the perfect hand to grasp as we gradually make this leap. Quantum art helps unhinge us from the default settings of our eyes, inner ears, and brain, the automated mechanisms within us that prevent us from seeing a physical world that consists of multiple and simultaneous physical realities.

Monet and Einstein were fixated on light, but this fascination was heightened by an even greater obsession with time. They both sought new ways to trace the effects of time by exploring the possibility of a more extended *now*. Einstein did this mathematically by calculating a stretched out "present experience" that would occur as a person approached the speed of light, or "c." Monet tried to represent the idea of an extended "present" by painting the same object over and over again in different light, subtly tracing the object's slight changes over time. By standing in front of a row of paintings of his Rouen Cathedrals, for instance, a person experiences the object, the cathedral in the "present" at the same time as he or she experiences the object changing slightly over time.

At the core of both Einstein's and Monet's (re)presentation of the present is light, or the placement of the sun. In fact, at the core of many of the principles of aesthetics and science is light and its effect on objects. Monet found ways to (re)present light in its many possibilities, in order to unlock our staid perceptions of reality. He was trying to capture time as it is etched in a longer, more drawn out *now* (the observer looking at one fixed object) as it changes across time.

Monet elaborated aesthetically what Einstein showed mathematically; that is, the notion of time is up for grabs. We can talk about "now" at "c" (at the speed of light) where "now" is a continuous event. The impressionist painters depicted physical reality in a way that disarticulates the most fixed and rigid notions of classical physics, thereby teaching us to see the world in its more elusive nature.

In *Art and Physics*, Shlain explores this process by very carefully deconstructing Édouard Manet's painting *Luncheon on the Grass*.[15] He shows how this work, without a word, rebukes classical physics and reveals a world unrelated to the traditional precepts of physical reality. He says "[t]he painting's inflammatory content and strange construction tacitly challenged Aristotle's logic and Euclid's space and called into question an entire paradigm built upon reason and perspective."[16] The horizon, in art as in our everyday lives, serves as an anchor against the reality that we are not at rest at all, but actually being hurled through space. Sailors focus on the horizon to avoid getting seasick; we focus on the horizon when we get carsick to gain a sense of our bearings, ground ourselves in something immobile, and draw the line *somewhere*. As Shlain points out, Manet moved the horizon up and up until it was finally off the page entirely. He says that "[b]y defiantly presenting arabesque verticals and a curved horizon, Manet challenged a mindset about space that had been born in antiquity and had remained essentially unchanged until it became petrified."[17]

Edgar Degas takes this a bit further by forfeiting both the vanishing point and the horizon. By doing so, he puts the viewer in a very precarious position that is outside of the strict parameters of classical mechanics. Linearity "grounds" us; gravity is comforting; a theory of absolute rest shows that we are still; dualism compartmentalizes the world; and Cartesian thought neatly separates us from that which the human mind cannot grasp. Clearly, Degas's work pulls up these anchors and inundates us with the beautiful unknowability of the world. Lines are blurred, while the nature of light is fluid, indeterminate, and larger than the mind can grasp. Shadow, perspective, and horizon are skewed and the chains of gravity are broken. The ethereal, complex, and layered ambiguity of nature shines through impressionist painting, thus initiating an evolution of consciousness and expansion of our understanding or "getting the world."

BEYOND "NEW PERSPECTIVE": ART AS CREATION OF A NEW REALITY THAT IS NOT REDUCTIONIST OR CAUSAL

The impressionists' paintings were the early twentieth century's kaleidoscopic models for the world; they provided a perspective on our environment that embraced relativity and uncertainty rather than an absolute and fixed viewpoint.[18] In the work of Manet, Degas, Monet, and others, we see the true nature of light and material reality, *impressions* of a reality that is changing, flickering, and neither staid nor determined.

Cubism took impressionism's cracked looking-glass perspective in a different direction, further distancing art from the rigid parameters of seventeenth-century philosophy and science. Beginning with Cézanne, the cubist painters performed alchemy with geometry and created a magical environment within which objects are animated, shifting, and moving off the page. In cubist paintings, we see consciousness extended to the material world via a kind of surreptitious placement of objects. The cubists' placement of objects skews the natural ocular perspective that our brains usually affix to what we see. This somehow lends a certain animated quality to *the material* and asserts the objects' unfixed place in curved space and relative time. In this respect, cubism is an extension of impressionism as it shows us that art can be more than simply an expression of form and color in space. Art *also presents a new reality entirely* that is neither causal nor reductionist.

The effects that this reinterpretation of the reality can have on accepted worldviews are staggering. The very precepts of seventeenth-century classical physics depended on the cohesion of the principles of determinism and duality, which, in turn, depend on Aristotelian deference to linearity and Euclidean geometry that, in turn, rely on noncurved space and fixed, rigid concepts of time. On a microlevel, Werner Heisenberg's Uncertainty Principle and, on a macrolevel, Einstein's relativity theories turn the ideas of duality and certainty inside out. They refute the very notion of a single frame of reference where a body sits in perfect stillness and enjoys a comprehensive "true view" of the world. The world, after all, is not at rest. Time and space are curved and relative, and the true nature of light cannot be perceived or measured at any given time. These new and fascinating precepts were beautified and rendered aesthetically interpretable by cubist painting.

The cubist painters reframed the very precepts of classical physics—space, time, and light—and geometry to create an entirely different worldview. While the impressionists sought to bleed the underlying reality of light through the overlaying existence of objects and make the former more prevalent than the latter, these artists were still more or less representing what was "seen" by the human eye. However, Cézanne sought to reconstruct the concepts of space and man's interaction with this dimension by rendering the notion of perspective relative and multitudinous.[19] One painting has several vantage points— that is, it reveals the several renditions of the space that one object occupies— with all of them true and viable. A piece of fruit is revealed simultaneously from several different angles. Objects are considered from many different perspectives at once, thus suggesting that objects do not exist simply as our mind sees them, but reside on several planes at once. Our minds have tamed this reality so that we only see one at a time. Cubist paintings help us to relearn to see the world in its more probable manifestations, rather than what our minds have taught us to apprehend. They represent the suspension of disbelief required by quantum physicists who must believe that, although we cannot measure it, the true nature of particles is varied.[20] In other words, reality can be "A" and "not-A" at the same time. Cubist paintings hold a window open to something that we could never see of our own accord; specifically, the qualities of physical reality (as quantum physics and cubist paintings have shown us) are not stable, predictable, or related by causality.

Niels Bohr, one of the founding fathers of quantum mechanics and what now encompasses, more generally, "quantum thinking," understood that the speculative nature of quantum physics was a necessary attribute of all quantum processes in general. Experiments continuously proved theories of quantum mechanics to be true, but nobody could explain how this occurred. This acceptance of unknowability as well as the belief in the existence of numerous and simultaneous possible outcomes, truths, waves, or, in other words, ways to see an apple are fundamental to quantum theory. This interpretation of the speculative nature of quantum theory came to be known as the "Copenhagen Interpretation of Quantum Mechanics."[21] This theory states that everything truly exists in a state that cannot be pinned down to one perspective at a time, and anything in its true nature is really a wash of possibilities, a vast sea of potential. The way we perceive reality is relative only to the way we frame or measure reality. In this way, we can look at a cubist painting as the oil-on-canvas version of the Copenhagen Interpretation of Quantum Mechanics. The implications of this parallel are enormous if you then think about the role of quantum art (art that elaborates on the quantum nature of matter and consciousness) in helping us literally to rewire our brains and see the world in a new, more expanded way. Quantum art trains our minds to see the more speculative, nondeterministic nature of the world.

Through impressionism, more through cubism, and later through abstract impressionism, we see that art history seems to be on an evolutionary path in that it increasingly represents a reality that is not reductionist or causal and that art also *creates* a world that is not reductionist or causal, a world where we can practice both seeing and "getting" reality. In quantum art, we can see the underlying motionless manifestation of an "inanimate object" when it is presented to us in a way that reaches beyond our mind's usual framework. The object expands our mind and lets us understand our environment as well as our own nature as multifarious, rather than exclusively uniform.

This is not a new idea that the precepts of seventeenth-century philosophy and science—determinism, causality, and reductionist thought—have been complicated by recent scientific discoveries and Eastern philosophy. Across disciplines, such as psychology, theology, and philosophy, we embrace more fluid concepts of reality and accept fuzzier contours of truth and experience. But ironically, the precepts of seventeenth-century thought, specifically the Newtonian worldview, still govern our most important societal and political structures. A lot of work needs to be done before we can fully embrace the more liberating concepts of new science and philosophy as part of the way we think about our every day lives. This is something that Danah Zohar expounds in her book *The Quantum Society*.[22] Zohar suggests that we think of the quantum aesthetic as a general philosophy and approach to studying social and physical realities so that we can begin to build a new framework for living similar to the way that Cézanne built a new framework for seeing. Our perception of art is tantamount to this larger effort. To think of the quantum aesthetics as a general philosophy and way of studying social and physical realities, the following should be established: (1) that there is a unified field that is probably a quantum system; (2) that consciousness is a dynamic part of this field; (3) that society in general is impacted by the expansion of individual consciousness, thus creating collective growth; and (4) that quantum art facilitates and accelerates this process of growth and expansion.

THERE IS A HIDDEN ORDER TO THE WORLD, AND IT IS PROBABLY A QUANTUM SYSTEM

In a recent episode of *The Simpsons*, an animated version of Stephen Hawking paid a visit to Springfield and mentioned that one of his regrets in life has been his inability to prove a unified field theory. The notion of a unified field, which would yoke relativity and quantum mechanics, apparently has found its place in popular culture. But beyond Springfield, scholars, poets, artists, and thinkers across disciplines have been enchanted by the idea that we all exist within a large, fluid matrix of some sort, a field of particles and waves that points to a hidden gel-like matrix, a sort of "jam of the universe."

Eastern thinkers, such as Buddhists, Shevites, Muslims, and Hindus are not newcomers to this notion. A "Oneness of Being"—a sense of a hidden source—is at the very center of Eastern belief systems. And in the past twenty years, a staggering number of Westerners have been drawn to these religious practices. Ashrams and Buddhist temples have become more prevalent in the West, as Westerners are increasingly drawn to the beauty of unity, quantum thinking, or a primary source that bleeds through the entire canvass of our lives.

Still, the recent enchantment with these ideas has deeper roots than popular television shows and the West's recent and avid adoption of Eastern belief systems. The dream of a unified field extends back in time before Socrates. Parmenides' notion of a "Oneness of Being" more or less described what Brian Greene, in *The Elegant Universe*, calls the unified field or a state in which "all of the seemingly different forces and permutations of matter in the world are really manifestations of one grand, underlying principle."[23] Parmenides referred to this underlying principle as the "One"; Heraclitus held a similar belief, but his underlying gel was called the "Cosmic Flux." Both believed that we are all made of one thing—the source. No matter what you want to call it, God, Prana, Chi, Tao, or Jell-O, for Heraclitus and Parmenides, we are a part of this origin in a manner similar to the way in which light is part of a flame.

Since the Newtonian worldview took hold in the seventeenth century, dualism and mechanistic thinking have become the backbone of modern civilization, and the "hidden order" has become more hidden still.[24] We are now making our collective way back to an understanding of our connection to this hidden order. For example, quantum field theory postulates a "vacuum." According to this theory, our universe consists of space, time, and the "vacuum." Even though this name implies emptiness, this vacuum is not empty at all, but rather represents the unseen potential for everything that exists. Think of the ocean; the white caps are everything that is manifest or visible, while wholly connected to and made of the underlying substance: the ocean. When you see a footprint in the sand, you see a few dots where the toes made their mark and a larger circle where the heel made an impression. Although the footprint may not be connected to anything that you can see, you know that this series of impressions represents a whole foot or something that is greater than the composite of these marks. In this same way, human beings, nature, and all of matter can be thought of as the visible marks that are parts of a larger, connected system or field that is hidden.

CONSCIOUSNESS IS PART OF THE QUANTUM SYSTEM

Many thinkers of recent decades have postulated that consciousness itself is a quantum system and that it too is part of the unseen order or field. John Wheeler, a student of Bohr, holds that the mind and universe are inextricably linked and that consciousness provides the link between the hidden order of the field, the behavior of electrons and quarks, and "everyday life." His Participatory Anthropic Principle tells us that the observer is necessary in the creation of the world.[25]

Zohar, in her books *The Quantum Self* and *The Quantum Society*, refers to the Bose-Einstein condensate as a possible model for consciousness as a quantum system.[26] The basic premise of this theory is grounded on condensed phase physics. In a quantum system, whose tiniest bits are interrelated like a hologram, a superfluid or superconductor material must be present in order for the interrelation to work. The Bose-Einstein condensate is:

> simply a system of vibrating, electrically charged molecules . . . into which energy is pumped. As they jiggle, the vibrating dipoles . . . emit electromagnetic vibrations . . . just like so many miniature radio transmitters. . . . Frohlich demonstrated that beyond a certain threshold, any additional energy pumped into the system causes its molecules to vibrate in unison. They do so increasingly until they pull themselves into the most ordered form of condensed phase possible—a Bose-Einstein condensate.[27]

We entertain this entirely speculative connection between the Bose-Einstein condensate and a model for consciousness by considering the tendency of molecules to "line up." When you consider meditation—or the slipping of mind, body, and spirit into "neutral—" you might think of this as a kind of "lining up" or coming into a phase of the layers of consciousness, so that individual consciousness becomes part of a larger quantum system or "field" that has characteristics of relational holism, as represented by a hologram.

SOCIETY IN GENERAL IS IMPACTED BY INDIVIDUAL GROWTH IN CONSCIOUSNESS

Teilhard de Chardin believed strongly that the expansion or development of consciousness experienced by one person results in a collective raising of consciousness. Shlain explains this process in this way: "Any time the consciousness of any one individual in the world is raised, the general quality and quantity of the mind in the world is enhanced."[28] Teilhard de Chardin called this invisible component of the atmosphere the "noosphere," after the archaic Greek word "noos," which means "mind." Each person, becoming aware of his or her life, adds to an ectoplasmic pool of awareness, thus ever so slightly raising its level.[29] At the center of his message is the belief that we are all linked by one macrosystem of consciousness that is affected by individual evolutions in awareness. This idea is a central premise of the quantum aesthetic as well.

INDIVIDUAL GROWTH AND CONSCIOUSNESS EXPANSION IS AFFECTED BY ART

So, what brings our consciousness "into phase," like the optimal state of a Bose-Einstein condensate, where the body, mind, and consciousness are acting in their optimal measure, where all systems are truly "go," and the connection between consciousness and field is most clear? Consider what Blake has said about art: Like Teilhard de Chardin, Blake saw the observer's interaction with art as something like meditation, and thus a means by which to reach a "higher state" or to "come into phase." With this model, persons' consciousness expands or evolves in a way when they "come into phase," meditate on a piece of art, or have a transforming aesthetic experience that promotes and undeniable "stirring." Teilhard de Chardin put it this way: "In short, art represents the area of furthest advance around man's growing energy, the area in which nascent truths condense, take on their first form, and become animate, before they are definitively formulated and assimilated. This is the effective function and role of art in the general economy of evolution."[30]

HEIDEGGER/VAN GOGH

Martin Heidegger, the German philosopher, believed in the ever-present mystery of the world (similar to Parmenides' "Oneness of Being" and Heraclitus' "Flux") and that certain works by particular artists act as catalysts in the process of "unconcealing" this hidden truth.[31] He was referring to what we are now calling quantum art. "Unconcealment," then, is the creation of a reality that uncovers a deeper, more whole, and universal truth that is neither reductionist nor causal. Heidegger believed that Vincent van Gogh's paintings provided a wonderful example of art's ability to illustrate the ineffable, unseen "mystery of the world." He said that there was a certain truth to van Gogh's paintings that rose out of the entire painting and was a product of the relational whole of the work. That is, beyond the object that van Gogh paints, there is a certain truth that permeates the painting.

Quantum art evokes perceptions about the inanimate and validates our inner, core intuitions about consciousness existing everywhere. Quantum art wants to look at the method of *unconcealment*, the revealing of what belongs to being itself. Quantum art allows us to "get" the myriad possibilities of the physical world without limiting ourselves to the one measurement or perspective the world is revealing to us at one, given moment in time. In this regard, quantum art upholds the coherence inherent in possibilities that are opposed.

BEAUTY AND ENTROPY—WHERE QUANTUM AESTHETICS FITS INTO THE SCHEME OF THINGS

The Second Law of Thermodynamics states that the nature of everything in the universe is entropic; in other words, all matter tends toward ever-increasing disorder. Entropy involves galaxies going slightly askew, orbits growing more unpredictable, desks getting disorganized, and piles of raked leaves spreading around. And yet, there is evidence of more harmonious laws at work that illustrate a tendency toward greater order and unity. Rivers run their steady course and resemble veins in their Earth-body despite entropy; molecules in a Bose-Einstein condensate come into phase; moreover, consciousness evolves despite entropy.

Is it necessarily a contradiction, then, that entropy may promote unity? Consider again Zohar's suggestion that the Bose-Einstein condensate might be a model for consciousness.[32] This Bose-Einstein condensate, a pumped system of electrically charged molecules, operates at optimum performance when inspired to do so, when some "excitation" causes molecules to line up, come into phase, and find order. This process, as should be noted, is the opposite of entropy. The mind and body work the same way. Studies show that, in a state of meditation, the human mind and body are in synchrony, at a calm hum, in their optimal, restful state. What is the impetus—the *excitations*—that causes the charged molecules in us, the universe, the field of consciousness, or a Bose-Einstein condensate to line up or come into phase? Remember what Blake said: Art is a catalyst for meditation and necessary to move forward on the path toward enlightenment. Could it be beauty, then, that has this profound affect?

Let us imagine it is something like "beauty" or, at the very least, aesthetics that influences consciousness in a manner similar to the way the electrically charged molecules of a Bose-Einstein system are excited and come into phase. This would mean that the "retinal experience" of an object's overall aesthetic properties, and the flow of energy around the object/environment in question, might have an influence on the consciousness (a pumped phase system) by promoting order in the environment and the beings that reside there. The basic principles of feng shui rely on this premise.[33] The Bose-Einstein condensate is a model for consciousness as well as for the effect of aesthetics on consciousness, for both processes illustrate, at least in part, that in consciousness there is a natural tendency away from entropy and toward order.

Quantum aesthetics *approaches* art as though this were true or, at the very least, possible. Aesthetics is seen as a concept that is tied to the physics of the universe as well as to consciousness. Quantum aesthetics sees art as part of a quantum system and playing a key role in our evolution toward order and connection. So, the quantum aesthetic does not represent an art movement that is new or unique, but rather an *approach* to art that combines concepts of physics and consciousness. As such, this outlook also refers to a way of perceiving art and its role in the expansion of the self and society.

The recent and increasing popularity and recognition of quantum physics, superstring theories, and a more expanded understanding of consciousness, along with their infiltration into the mainstream, speak to what Teilhard de Chardin refers to as a "tightening beyond ourselves and above our heads, of a sort of cosmic vortex, which, . . . pushes further . . . towards a continuous and simultaneous increase of complexity and consciousness."[34] The recent momentum behind quantum aesthetics is part of the same "evolution." So how do we arrive at the notion of a quantum aesthetic at this point in our intellectual/art/science/spiritual history, and why is it gaining such momentum now?

In the past twenty years, we have witnessed the death (on an intellectual/cultural level) of the Newtonian/Cartesian paradigm at the hands of postmodernism. Postmodernism was quickly followed by a movement called avant pop, a kind of postmortem reevaluation and looting of the wreck that was postmodernism.[35] Now, we are yearning for a deeper, more meaningful connection to ourselves and to art, one that avant pop cannot provide. This is where the quantum aesthetic fits in.

POSTMODERNISM

Postmodernism rang the final death knell for the precepts of modern thinking, which had its roots in seventeenth-century physics and philosophy. By the 1980s, postmodern and deconstructionist thinkers such as Jacques Lacan, Michel Foucault, Jacques Derrida, and Paul de Man had taught us how to second-guess the meaning of language. Forfeiting the connection of the signifier to the signified and uprooting the very linear language system that Shlain holds was the precursor of all dualist thinking, postmodernism and deconstructionism lifted the final intellectual anchors of seventeenth-century worldviews.[36]

In the heyday of postmodernism, university students of literature and philosophy arrived at their lecture halls to find language stripped down and rendered meaningless. The narrative form was highly suspect and dangerous, along with notions of gender or "universal truths." God was dead, and so was any kind of author(ity). Consider high modernist culture thought to be a glass house and postmodernism the house after an earthquake: apocalypse, entropy.

If postmodernism did kill philosophy, language, the author, and God, then, to what end? Picture this crime scene: Glistening shards lying strewn on the floor, broken images reflecting useless slivers of information, and images broken out of context and rendered totally meaningless. The only real truth or meaning in sight is the shock value of the apocalypse itself. Consider Thomas Pynchon's novels. Like in Francis Bacon's paintings, the suspended moment of horror—the captured moment of utter alienation—is a quintessential image of the postmodern condition.[37] The postmodern image of the self is fragmented, shattered, and the end result of entropy.

Despite the successful breaking of ties with oppressive seventeenth-century-to-modern worldviews, the postmodern condition left us in need of some kind of reconciliation. If postmodernism was a glimpse of a belief system in entropy, an intellectual apocalypse, then avant pop is a response to that, an intellectual/artistic reaching out for reconnection, order, and unity.

NOSTALGIA AND LONGING: AVANT POP

Avant pop was the name of an album by 1980s jazz musician Lester Bowie. Bowie combined melodies of popular songs like "Blueberry Hill" with riffs of his own creation.[38] Avant pop came to be associated with an art and literature movement that is characterized by references to popular culture (songs, ad campaigns, world news events, television stars, and tennis shoes) that tap into a collective popular memory filled with trivial, useless information. If postmodernism destroyed the glass house of modern thinking, avant pop helped us rebuild it by tapping into a collective nostalgia that pieces the shards back together, finding order and connection by using the very fallout of the apocalypse. Avant pop rearranges into a clever montage the signifiers of popular culture to create a new sense of meaning, which we can relate to because of our shared knowledge of popular culture. If I say "Bruno Magli loafers," you think of O. J. Simpson. If I write a short story whose main character is called Niles Crane, I can rely on your familiarity with the character Niles Crane of the television show *Frasier* to inform my character's personality. I can assume that we have these themes in common through popular culture.

Presaging this movement, and always ahead of his time, was Marcel Duchamp. His so-called ready-mades were artifacts of the everyday, for example, bicycle wheels and toilets, that combined in a novel way and given new meaning.[39] Recognizing an everyday object, then seeing it in a new, retooled way was the essence of avant pop art. Musically, avant pop is Beck Hansen. "Beck," the flamboyant pop star from Los Angeles, is famous for combining reminiscent musical riffs and jingles into songs that become new and all his own. His trampoline lyrics make reference to everything from brand names and blue jeans to Duchamp himself, in a Gertrude Steinesque–style that he has made into avant pop anthems. Beck draws on media culture nostalgia by tapping into our collective memory—that is, the vast storage units of commercial jingles, television theme songs, designer jeans, and the emotional connection we have to Jordache, Sergio, Levis, Kraft Macaroni and Cheese, TV dinners, the cool blue pool of light in front of a television set, and the luminous halo of our media culture. At least in the United States, avant pop promotes the extreme of "empty paper cup (Dixie) culture" and establishes this framework as a unifying social force.[40]

How, then, does avant pop set the stage for quantum aesthetics? No matter how many times you recombine empty, essentially meaningless elements, the end result still yields nothing. The nostalgia and longing for a connection to order that occasioned avant pop cannot be satisfied by the meaningless jingles and references to popular culture that are the product of avant pop art. In this sense, quantum aesthetics picks up where avant pop leaves off. Avant pop declares a belief in meaning, no matter how shallow the ad tag line or empty the commercial jingle; avant pop asserts a tendency toward connection, or some kind of unification. The quantum aesthetic responds to this need for connection in a way that avant pop cannot.

The quantum aesthetic is an approach to art, a way of seeing *in* art our deep connection to the universe. Quantum art brings us closer to that union and helps us see it more vividly. Through the aesthetic experience of quantum art, we become more connected to what we know on an inner, intuitive level to be true.

Socrates said that our soul knows everything already. The quantum aesthetic is an answer to our innate longing to be in touch with our knowing soul. Rumi explained this universal longing in this way: As soon as we are born into this world, we are separated from our true source. We spend our lives trying to find our way back, in a sense, looking for clues that we do belong to a larger order. We pass on stories of amazing coincidences and miracles to fortify our faith in a larger, more beautiful rationale for the world. This gesture calls to the soul so that we might feel deep within ourselves a stirring response. Quantum aesthetics is an approach to art that teaches us how to feel that stirring through observing more expansively the things that surround us and to create art in such a way that we might become better attuned to the soul's deeper connection to the world.

THE SECRET LIFE OF BREAD

René Magritte, considered one of the most important surrealist painters, differed from his surrealist colleagues in a very important way. As opposed to Salvador Dalí and others, Magritte did not paint in the surrealist mode in order to plumb the depths of his subconscious or to unearth images and archetypes buried deeply within him. Instead, he painted scenes that would challenge and stretch the abilities and dimensions of his lucid, conscious mind. Through the experience of painting and reflecting on his own paintings, he sought to expand his conscious awareness and to "get" his experience of the world as limitless.[41] Magritte's act of painting, then, is a gesture that validates all that the quantum aesthetic suggests; that is, art expands us, and as we expand on a conscious level, so does the world.

At a recent Magritte exhibit at the San Francisco Museum of Modern Art, I found myself drawn to the painting, *The Golden Legend*. It depicts a half dozen baguettes floating past a window on a still, indigo, and starry night. I sat in front of the beautifully strange painting for a long time, while the all-too-many Thursday night museum-goers shuffled by in gray shoes and black overcoats. I knew I was drawn to the enchanting suggestion of the painting, the notion that bread might have a secret life of its own; beyond that, I could not explain my fascination or my need to spend as much time as possible in front of this painting.

I also knew that the reason why I was drawn to the painting didn't matter as much as my simple, quiet reverence for the exchange that was taking place between my body/soul/mind and the painting itself. It's these moments, according to the quantum aesthetic, that act on us, that change us, and that have the potential to alter the evolutionary path of consciousness that we all participate in.

The precepts of the quantum aesthetic are revolutionary because they suggest that such a process (art influencing the evolution of a universal consciousness) is possible. And yet, since the industrial revolution, visionaries like Blake began to pull away the cobwebs and shackles of seventeenth-century thought and artists began liberating the more flexible and beautiful truths hidden by mechanism and deterministic thinking, thus paving the way for more artists to help guide us away from mechanistic, dualist thought. Impressionists and cubists prepared us for the astounding discoveries of new science and quantum mechanics. And our popular culture, through the avant pop movement, allowed us to bounce back from and heal the wounds caused by postmodernism's wrecking-ball effect on meaning and reality.

The quantum aesthetics, as described by Gregorio Morales and the many writers and artists who embrace its theories, is finding in the new millennium a population ready to accept the consilience of art, literature, physics, and consciousness. It's not a coincidence that art history, scientific discoveries, and a theory like quantum aesthetics are converging at one time. Our ability to see the world is evolving on a universal scale toward a more complete acceptance of the universe's "hidden meanings."

So we continue to pass on stories of amazing coincidences and uncanny parallels between the fields of science, art, and philosophy as way of giving credence to our core intuition that the universe and all that exists within it is a beautiful tapestry of meaning, a dance that is self-choreographed. Quantum aesthetics provides the perspective we need to continue to expand our ability to grasp this understanding of the world and to more consciously participate in the evolution of a universal consciousness.

NOTES

1. Leonard Shlain, *Art and Physics: Parallel Visions in Space, Time, and Light* (New York: Morrow, 1991), 383.
2. Katherine Hoffman, *Explorations: The Visual Arts since 1945* (New York: Icon Editions, 1991), 21–23. Mark Rothko was one of the main painters of the "color field" movement, which grew out of abstract expressionism in the 1950s. Rothko's paintings are easily identifiable by their large blocks of ethereal color that seem to float in a state of mesmerizing translucency.

3. This interview took place in San Francisco's Herbst Theatre on Monday, March 6, 2000, as part of a series of interviews hosted by *City Arts and Lectures* of San Francisco.

4. Bruce Cole, and Adelheid Gealt, *Art of the Western World: From Ancient Greece to Postmodernism* (New York: Summit, 1989), 243–247.

5. Alan Lightman, "Relativity and the Cosmos," *Einstein Revealed*, 22 September 2000 ‹http://www.pbs.org/wgbh/nova/einstein/relativity/index.html›.

6. Pierre Teilhard de Chardin, *Building the Earth* (New York: Discuss, 1965), 22–25.

7. Gregorio Morales, *El Cadáver de Balzac* (Alicante: Epígono, 1998), 14–35. This first chapter of Morales' seminal book outlines the premises of quantum aesthetics theory.

8. Shlain, *Art and Physics*.

9. Ibid., 75.

10. Ibid.

11. René Descartes, *Selected Philosophical Writings* (Cambridge: Cambridge University Press, 1988), 143–150.

12. Geoffrey Bloom and Erdman David, eds. *The Complete Poetry and Prose of William Blake* (Berkeley: University of California Press, 1982), 82.

13. William Blake, *The Marriage of Heaven and Hell* (New York: Granary, 1993).

14. Brian Greene, *The Elegant Universe: Superstrings, Hidden Dimensions, and the Quest for the Ultimate Theory* (New York: Norton, 1999), 97–105. In this section of his book, Greene explains the many experiments that have led to the fundamental belief that light exhibits both wave and particle properties and that accepting the duality of light principle is fundamental to understanding quantum mechanics. He says: "[L]ight has *both wave-like and particle-like properties*. The microscopic world demands that we shed our intuition that something is either a wave or a particle and embrace the possibility that it is *both*."

15. Shlain, *Art and Physics*, 102.

16. Ibid., 104.

17. Ibid., 106.

18. Of course, it can be said that Georges Seurat and other pointillists tried to reduce reality to its basic elements, or dots. I would argue that the end result of the pointillists' work was not a reduction of reality to simple dots, but rather another gesture toward revealing the elusive "true" nature of light and color as more significant to our sense of reality than definitive lines, shapes, the absolute separation of objects, and the space that exists within.

19. Cole and Gealt, *Art of the Western World*, 254–255.

20. Morales, *El Cadaver de Balzac*, 27. Morales points out that Albert Kosko's principle of fuzzy thinking is a cornerstone to the quantum aesthetics. See Bart Kosko, *Fuzzy Thinking: The New Sciences of Fuzzy Logic* (New York: Hyperion, 1993). "Fuzzy thinking" can best be described by the axiom that "A" and "not-A" can exist simultaneously. In fact, "A" can equal "not-A." For instance, light exhibits both wave and particle properties. Though this cannot be necessarily explained, experiment after experiment shows this to be true. It's fundamental that quantum thinkers remain open to the possibility that the essential substances of the universe can exhibit opposing characteristic. This is also the premise of the Copenhagen Interpretation of Quantum Mechanics.

21. Wheeler, John Archibald, *Geons, Black Holes & Quantum Foam: A Life in Physics* (New York: Norton, 1998), pp. 123–124. John Wheeler explains that Bohr's Copenhagen Interpretation of Quantum Mechanics was extremely controversial. The Copenhagen Interpretation relies on uncertainty, speculation, and probability to predict outcomes of quantum events.

22. Danah Zohar, *The Quantum Society: Mind, Physics, and a New Social Vision* (New York: Quill, 1994).

23. Greene, *Elegant Universe*, 4.

24. David Bohm, *Wholeness and the Implicate Order* (London: Routledge and Keagan Paul, 1980).

25. John Barrow and Frank Tipler, *The Anthropic Cosmological Principle* (Oxford: Oxford University Press, 1988), Chapter 10. Wheeler's Participatory Anthropic Principle holds that consciousness operates within the greater universe as part of a quantum system. Therefore, just like in any quantum system, each element has some effect on all other elements in the system. In this way, the "observer" acts as creator in the physical world.

26. Zohar, *Quantum Society*, 78.

27. Danah Zohar, *The Quantum Self: Human Nature and Consciousness Defined by the New Physics* (New York: Quill, 1990), 83.

28. Shlain, *Art and Physics*, 383.

29. Ibid.

30. Ibid., 387.

31. Martin Heidegger, *The Piety of Thinking: Essays by Martin Heidegger* (Ontario, Canada: Indiana University Press, 1976), 35–39.

32. Zohar, *Quantum Society*, 78.

33. Kwan Lau, *Feng Shui for Today* (New York: Tengu, 1996), 11. Feng shui is the ancient Chinese practice of placing objects in specific places within a space in order to improve the harmony of the space, as well as the health and general well-being of the people who spend time in that space.

34. Shlain, *Art and Physics*, 383.

35. Larry McCaffery, ed., *After Yesterday's Crash: The Avant-Pop Anthology* (New York: Penguin, 1995), xv–xvi. In this collection of avant pop literature, McCaffery explains that the advent of the avant pop movement was occasioned by postmodernism's destruction of high modernism. He says, "[a]ccording to the leaders of the PoMod Squad, not only had serious art died but so had a lot of other things—including meaning, truth, originality, the author (and authority generally), realism, even reality itself." Meanwhile, "a new breed of media-savvy American writers and artists were busy down in their basement laboratories mixing up some new kinds of aesthetic medicine, specifically designed to revitalize artists suffering from info-overload, psychic fragmentation, loss of affect, reality, decay, daydream drift, and other debilitating symptoms of life in post-apocalypse America."

36. Shlain, *Art and Physics*, 17–18.

37. Thomas Pynchon, considered to be the quintessential fiction writer of the postmodern era, is famous for vividly depicting the postmodern condition. His novel *Gravity's Rainbow* (New York: Viking, 1972), with its nonlinear narrative, shifting perspectives, constant images of loss, and destruction and alienation, presents a horrifying picture of reality that has become recognizable as the "postapocalyptic postmodern condition."

38. McCaffery, *After Yesterday's Crash*, xx.
39. In Cole and Gealt, *Art of the Western World*, 273–274.
40. McCaffery, *After Yesterday's Crash*, xxvii. McCaffery describes "Avant Pop's depiction of an info-overloaded, remote-control culture capable of accessing innumerable realities with just a casual flick of the joy stick." The culture that avant pop propagates is best described by its fiction, according to McCaffery. With respect to avant pop fiction, McCaffery says, "[a]s the media's scope seems at once to expand outward and advance inward, it becomes increasingly difficult for characters to be able to isolate an irreducible authentic 'me' that can be separated from the constructed desires, memories, and opinions occupying their minds." The "culture" that avant pop reflects is media culture. It is at once as widespread and unifying as any number of Eastern or Western religions and as shallow and temporary as a commercial slogan.
41. Cole and Gealt, *Art of the Western World*, 288.

CHAPTER 6

Quantum Aesthetics: Art and Physics[1]

María Caro and Andrés Monteagudo

> When the vision of the revolutionary artist . . . combines with precognition, art will prophesy the future conception of reality. The artist introduces a new way to see the world, then the physicist formulates a new way to think about the world. Only later do the other members of the civilization incorporate this novel view into all aspects of the culture.
>
> Leonard Shlain[2]

The Quantum Aesthetics Group, as it is now known, was formed in Granada during the winter of 1999 around several artists' and writers' common interest in certain scientific issues related to their respective disciplines. However, a few months earlier, the group started to be conceived under much more confusing and ambivalent names and objectives. "Contemporary Mystics" or "Quantum Mystics" were two of the names thought of to refer to the type of knowledge that lets us get to the total reality and/or the entirety of the universe. Indeed, with that meaning of the term "mystic," we intended to leave out any form of theism.

At that point, we already thought that the interrelationship between human beings and nature was crucial for our understanding of the world. We believed in the existence of different states and levels of consciousness in everything that surrounds us, and advocated for a humanitarian and ecological thrust in our group. We wanted art that sought the active participation of the observer, catalyzed intimate communication and dialogue among reviewers, and, in general, produced a positive personal reaction.

As we discovered the principles of quantum physics, we realized that they encompassed everything we thought to be important and opened a wide investigative field with different expressive paths and possibilities. We could not ignore the fact that quantum physics had changed the way reality is seen and understood. Accordingly, from our viewpoint, art could not remain oblivious to these changes. It was then that the artist Xaverio, through art critic Juan Carlos Martínez Manzano, contacted the writer Gregorio Morales, who had been working with these ideas longer than anyone else has. In his book *El Cadáver de Balzac*, he mentioned some of the theses on which we based our artistic work. At the meetings that we had regularly with Morales's literary group (Fernando de Villena, Francisco Plata, Miguel Ángel Contreras, and so on), the theoretical bases of the Quantum Aesthetics Group became increasingly refined. In La Zubia, a little village near Granada, the group formally took form and its *E-m@ilfesto*[3] was signed.

In this document, the point is made that the "*Quantum Aesthetics Group* intends to explore the creative avenues open by the most advanced science and psychology, turning art and literature into precise instruments capable of inquiring about human complexity and everything that surrounds humankind."[4] Thus, the formation of the artistic group was intended to unite artists who work under similar premises, but do not necessarily share similar techniques or interests. What is distinctive about this group is the variety of styles, materials, and mediums that the artists use to create their work. This diversity is enriching, opens doors for continuous dialogue and different evolutions, and mirrors the diversity that exists in contemporary art.

Therefore, as a nondogmatic group, we do not advocate a formal revolution in—a rupture—or a disintegration of the current art world. We simply open ourselves to knowledge, specifically to that which, coming from physics and mathematics, helps us establish a different artistic path with important connections to scientific advances—at a moment when science connects more and more to the humanities.

Due to time and space constraints, in this chapter we only briefly discuss the personal ways in which the different artists interpret the ideas and premises of quantum aesthetics. We also provide examples of their work and analyze the theory behind them. But first, we need to stop for a minute and discuss the relationship that we see between art and science, for our works are created in terms of this relationship.

Our *E-m@ilfesto* says that "*Quantum Aesthetics* entails the creation of both an art and a literature based on the most revolutionary findings within quantum physics and the psychology developed after it."[5] Thus, it is important for the artists in the Quantum Aesthetics Group to study the history of philosophy, art, and science so that we can understand the reasons why quantum physics became important and capable of influencing art. In the same way, it is important for us to study and compare some of the elements that both scientists and artists use in their work. We believe that while artists and scientists approach their interests and present the results of their work in a different manner, their basic questions and answers are quite similar.

Quantum artists are neither the only nor the first ones who have based their work on physics or mathematics. Matila C. Ghyka, in his *Estética de las Proporciones en la Naturaleza y en las Artes* (*Aesthetics of Proportions in Nature and Art*)[6]—a product of the interest that the discovery of Renaissance treaties about perspective, mathematics, and architecture provoked in the author—analyzes the relationship between art and science at different times throughout history. Ghyka discusses the Egyptian studies of proportions, Parmenides' geometry, Plato's divine golden number, Leonardo da Vinci's treaties, and the various modern approaches to science. In this historical journey, he emphasizes the ongoing relationship between science and art: both of them search for, in nature and its creatures, the general mechanisms of a universal quest for knowledge.

However, as Ernst Cassirer says in his *Folosofía de las Formas Simbólicas* (*The Metaphisics of Symbolic Forms*),[7] although Parmenides denied that "pure knowledge" had any contact with nonbeing, science has tried to maintain such a separation. A good example of this is the way in which positivism declares that science can only come to fruition by jettisoning all of its mythical and metaphysical components. However, after studying positive science thoroughly, we can see how these metaphysical factors that were thought to be overcome by science still lay at its basis. For example, Auguste Comte, the most important figure of nineteenth-century positivism, ends by subsuming his positive science under a mythical and religious structure.

Nevertheless, this positivistic approach starts to crumble following the new discoveries in mathematics and logic. Furthermore, the unsettling findings of quantum physics have changed the Newtonian conception of the universe: space, matter, and time seem not to have clear-cut boundaries any longer,[8] logic becomes fuzzy,[9] and light is both a particle and a wave. Since the beginning of the twentieth century, science began to abandon its dualism and radical positivism in favor of Werner Heisenberg's idea of indeterminism. It became clear that we cannot have access to reality without the filters that are vital to human beings and their measurement gadgets, that is, their senses. Reality will always be mediated by these senses and thus cannot be singular but diverse, cannot be universal but personal, and cannot be located outside but inside the individual. Now we cannot talk about the truthfulness or falsity of a theory in general terms anymore, but only within a limited context. In this way, the usefulness of any knowledge begins to take priority over its veracity.

Truthfulness, in this way, becomes less important and the ineffable takes a paramount position. Both poetry and art have began to recognize this change. Currently, it is not rare to find thinkers in various disciplines who use intuition and fantasy as the bases of their work and as a mode of inquiry. For example, Juan Pérez Mercader, a well-reputed astronomer, uses intuition, as well as mathematics and numbers, to explain the universe. Intuition takes us to unreachable and intangible forms of nature and allows persons to expand their minds and their personal and collective lives. Additionally, intuition provides the capability of entering into the "morphogenic fields," within which the information that connects each individual to all past individuals resides and is transmitted through time and space.[10] However, imagination, a factor symbiotically related to intuition, takes us to levels of creativity that help us search deeply into the complexity of our minds. As a result, both fantasy and intuition, which have been despised by science but loved by art, become central to developing a holistic theory. Without these two characteristics of the human being, no scientific hypothesis can encompass the whole of reality.

Two other elements have been condemned by science in order to preserve objectivity but have been praised by art: myth and symbol. Symbols create connections to the realm of the metaphysical; they provide links between the individual and his/her exterior, intuitive projections of both the universe and cosmos that extend the human being in the search for absolute knowledge and renewed consciousness. This consciousness is conditioned by an awareness of nature and the world, as well as by insights into the self. According to Cassirer, "the world of myth and symbolic thinking is a fact similar somehow to the world of theoretical knowledge."[11] In this way, Cassirer acknowledges that both mythological and symbolic thoughts are valid forms of knowledge equal to that of other disciplines. Cassirer goes on to say that:

The images of the cosmos, of the cosmic space, and of the distribution of bodies in space that astronomy gives us, were originated by the astrological intuition of space and access to this space. Before being pure mechanics and a mathematical expression of the phenomenon of movement, the general theory of movement tried to answer the question about the 'source' of movement, which dates back to the mythical problem of creation, of the 'first motor.' And no less than space and time, the concept of number, before being a pure mathematical concept, was a mythical concept. . . . Before it was a measurement unit, the number was worshiped as a 'sacred number.'[12]

So, mythical space occupies an intermediate position between sensitive perception and pure knowledge—the space of geometric intuition. In this regard, space, time, and number are ideal factors—as ideal as intuition or fantasy—that intervene in the general task of attaining knowledge.

In the work of Andrés Monteagudo, a member of the Quantum Aesthetics Group, some of the aforementioned elements are paramount. From the very beginning, Monteagudo talks about art as a science and science as a form of art, and uses both as ways of observing nature and the universe and inquiring into the original reasons that sustain the individual's internal and external worlds. In his work, Monteagudo uses intuition and imagination—the very elements that scientists, philosophers, and artists use to create their work and transcend reality and to make existence more understandable.

As a result of these theoretical bases, Monteagudo understands his art and his way of conceptualizing art as a nonmystical ritual, an "anthropological rite," and a meeting or a starting point for speculation. Through this ritual, Monteagudo seeks and emphasizes forms that are present in the universe. In this way, the author participates in the connection between the physiological, metric, and mental space on the one hand, and on the other the geometric intuition found in mathematics, physics, and aesthetics. In numerous personal conversations, Monteagudo tells us that "mathematical thought and quantum physics, which prescribe certain conclusions about the physical world and its relationship to humanity," make him reflect "on the relationship between the two types of space: internal (that which our mind gives us) and external space (that which gives us a position and a direction, such as right, left, up, and down)."[13] This interrelationship between spaces has characterized the work of Monteagudo from the beginning. This is the reason why he is very interested in the quantum concept of vacuum; a vacuum that contains a great amount of energy—zero-point energy[14]—that is generated by the constant and haphazard production and destruction of virtual particles. But also, this is a vacuum that is filled with the individual who observes and feels space, a vacuum before which the individual must position him/herself. This vacuum, in sum, is not really empty.

Monteagudo regards the theories of quantum and subatomic physics as very impressive, for they specify that time and space are filters that the individual employs to create his/her own reality. No scientist can nowadays deny that aesthetics, which is the product of our mind's internal space (as is mathematics), provides us with the ladder necessary to see the entire universe.

Monteagudo bases his installation "Habitáculos" ("Dwellings") on these ideas (Photos 6.1 and 6.2). In this installation, he uses the human body and its different habitats as yardsticks to measure the universal environment that is space. Monteagudo wants to discover new places by using the directions given by the marginal lines of architecture and corners that serve as the intimate and private shelters that constitute our reality. Through the angles of these habitats, our gaze unconsciously provokes an escape toward a different place where our mind can either be a part of space or establish the dimensions of space. In this place, we can locate matter and the intersection between the vertical and horizontal, and also create a map with which to move around the maze that is constituted by the different levels of our mind. We can also try to find the last corner, the last edge that takes us to what all disciplines intend to accomplish: knowledge of the whole.

In addition to all of the theories discussed so far about the connection between art and science—particularly the relationship between thought and reality, intuition and number, or myth and science—there are some other tenets of the Quantum Aesthetics Group that we take very seriously. After all, they are compatible with the parameters within which our work is created. Following the principle of nonseparability, we understand, for example, that the cosmos is a fluid in which everything is interconnected: matter and consciousness are part of the same magma.

Photo 6.1. Andrés Monteagudo. "Habitáculos" ("Dwellings"), Installation, 2000.

Accordingly, it is important to note the "sacred" character that matter has for the members of the group. According to Mircea Eliade, "men keep participating in the sacred with certain behaviors such as their love for nature. The sacred survives wrapped in the unconscious, and humans, after all, keep feeding on the unconscious and let themselves be guided by this facet of the mind."[15] To be sure, Nature (matter) is revered by our group, for we feel that it possesses levels of consciousness that are related to our deepest self, is part of us as much as we are part of it, and also has the sublime quality of surviving us in time.

We see these qualities reflected in the work of Agustín Ruiz de Almodóvar (Photos 6.3 and 6.4). This author attributes to matter (mud and ceramics) human characteristics such as consciousness and a soul. This is why his whole

Photo 6.2. Andrés Monteagudo. Detail of "Habitáculos," Installation, 2000.

work points toward matter. He configures matter in architectural ways by limiting emptiness with ordered and geometric sets of "solid windows, opaque glasses, and permeable walls."[16] He uses these contradictions as a metaphor of the self in which "A" and "not-A" can exist together at the same time and in which the "integration of opposites into the whole"[17] can take place. His work is thus a human architecture that confronts observers and in which observers recognize themselves; an architecture around which observers wander and whose interior observers penetrate.

The constitutive elements of Almodóvar's art are emptiness and fullness, matter and nothingness, and memory and forgetfulness. With these components, he takes us to the time when men sought their transcendence by leaving their trace on matter, and thus participating in nature's fundamental thrust: perdurability.

Photo 6.3. Agustín Ruiz de Almodóvar. Untitled, 1998. Used with permission.

Photo 6.4. Agustín Ruiz de Almodóvar. Untitled, 1998. Used with permission.

Xaverio's work has also been designed to understand and express transcendence. Throughout art history, we have found recurrently the artist's intention of transcending reality through his/her art. Thus, we have to acknowledge the power of art to transcend appearances and plumb the depths of reality. Many great works throughout art history were created with this intention in mind; sometimes this process took the form of mystical transcendence and at others it pertained to revealing the unknown. Through conversations with Xaverio, we have learned about the importance he gives to seeking this transcendence and discovering what is hidden behind the appearances of daily existence. These are the reasons why he tries to find similarities between oriental thought—for example, meditation techniques, assimilation to the environment, and knowledge about mind, matter, energy, and emptiness—and those quantum theories that deeply transform the way in which we understand ourselves and our role in the universe.

Xaverio has made different serial works inspired by quantum theories, such as the wave-corpuscle duality, the implicate order, the correlation among particles, the vacuum, and the spinning of subatomic particles. The results are works that lead us to different conclusions depending on the theory on which he chooses to elaborate. Some of his works, predicated on both how the smallest thing is capable of representing the biggest entity—the microcosmos expresses the macrocosmos—and the wave-corpuscle duality of light, are sometimes perceived as light and at others as matter. Furthermore, some other works can be appreciated in any position in which they are placed—Xaverio has named this quality "quantum position." Yet, other works need some time to be apprehended and all the senses to be captured (sight, hearing, taste, touch, and smell). This is the case of Xaverio's installation, "Siesta" ("Nap").

In his "petrales" (works made out of crystals, gems, and minerals), observers let their gaze rest on a particular surface, and after closing their eyes they acquire an internal image that is called an "after image" or "captive image." Through this image, the subject is reflected in the object; the subject both is in the object and contains the object. Hence, through this image the integration of mind and space, or inside and outside, is achieved.

In his installation "Colores para Pasear" ("Walking Colors, or Colors for Strolling"), Xaverio has created a work that is longer than thirteen feet, around which the observer walks. This work lets the observer enter a sort of virtual reality where color is a very disturbing and unsettling factor, because each color is related to a state of mind. After all, Xaverio has been studying for quite some time how light reflection and refraction on matter produces special conditions that can alter the ways in which an artwork is observed.

The Swedish artist Mikael Fagerlund shares with Xaverio an interest in both the surprising aspects of light's behavior and the way the human eye captures different sensations depending on how the individual interacts with light. Fagerlund breaks down the logic of vision by developing the surface of a painting in such a way that three dimensionality becomes flat. The eye is trapped between the senses and reason, between passion and pure knowledge. Therefore, the eye must integrate both poles and break with the dualism of Newtonian science, for the only way to apprehend reality, as revealed in Fagerlund's work, is by experiencing the world through reason and reasoning through feelings. This is how observers become masters of their gazes, choose their spaces—real or imaginary—and produce their final works. With these ideas, Fagerlund proclaims a revolution in our sight: within the eye lies a possibility for cultural liberation.[18] (See Photos 6.5 and 6.6.)

Science and fantasy are mixed in Fagerlund's installations, sculptures, and paintings. But, as the artist tells us, this science is liberated from the burden of developing universal theories and rigid predictions and is endowed with the freedom of exploration and expression. Fagerlund's science is artistic and extends beyond what the eye can see.

Another tenet of the *E-m@ilfesto* that is implemented by the artists of our group refers to the human being as an active "imaginator" of the universe and creator of his/her own reality. This idea is used actively by María Caro in her work. For her, quantum physics opens up the possibility that different realities—one for each person—exist. Quantum physics even opens up the possibility of a reality that is not based on matter alone. After Heisenberg proposed his indeterminacy principle, the emphasis of any explanation must be based on humankind, on what individuals see, hear, and feel, for only through these means is reality grasped. Humans are both the measurement tools and the filters of all data, as David Deutsch tells us in his essay on the structure of reality:

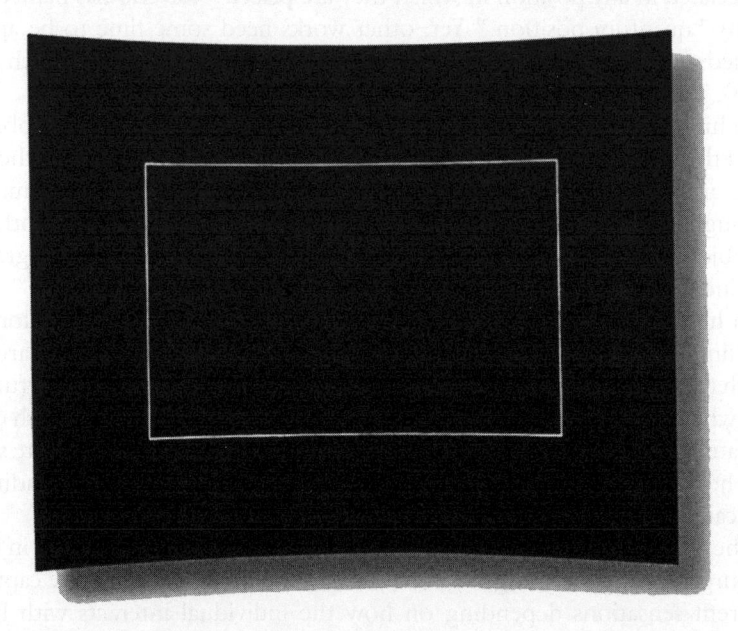

Photo 6.5. Mikael Fagerlund. "White Rectangle on Blue Concave," 1997. Used with permission.

Photo 6.6. Mikael Fagerlund. "White Rectangle on Blue Convex," 1997. Used with permission.

"After all, the solution to any problem takes place inside the human mind."[19] Thus, according to Caro, "art is ahead of science in that, since the very first artwork, the inside of the human mind has been the basis of artistic creation. We artists are used to treating the ineffable as the only reality."[20]

Since she started her artistic career, Caro has worked with series and the repetition of modules in such a way that the part is both finished and complete—for it contains the whole—and a component of an unfinished whole. We could say that Caro is creating just one artwork that is always the same. This single piece, as a whole, constitutes what she calls "the garden of Aleph"[21]: something that the human being creates as a lived-space with Nature as the raw material. This is a place where everything is potentially contained, and thus where the individual, as he/she evolves, discovers everything that already exists.

As we have mentioned throughout this chapter, after the quantum revolution, science turns to the *inside* because this parameter is necessary in order to achieve a holistic explanation of the world. Indeed, Caro is interested in this *inside*. Better yet, Caro is interested in the exterior of our minds that is filtered by our cognition. Caro's works are produced in this process; they start with interiority and open up to the outside, where Caro asks for an intimate answer. Caro's work originates from the confrontation between the inside and outside; this dialectic is the raison d'être of her art. Her installation, "Jardín" ("Garden"), for example, intends to create an internal locale that is a metaphor of a garden made according to to our ruminations, memories, and fantasies (Photos 6.7 and 6.8). But this place is also a deep space of bright journeys and

Photo 6.7. María Caro. "Jardín" ("Garden"), Installation, 2000.

shady grottos that nobody can enter. After all, "Jardín" recreates a type of architecture of the soul, an internal scenario that needs to provoke an external event. Without such an event, which occurs within the observer, the artwork would be unfinished.

The concept of sensation—unique and unrepeatable—that makes the human being an essential part of the universe, attracted all of us, as artists, from the very beginning: the individual is both a whole and part of a bigger whole. Correspondingly, the idea of individuation that Gregorio Morales presents in *El Cadáver de Balzac* became very important for us. According to Morales, "individuation is the task of extracting what is unique from what is common; differentiating what is ours from what has been imposed on us by the commu-

Photo 6.8. María Caro. Detail of "Jardin" ("Garden"), Installation, 2000.

nity. . . . In Jung's words, there is a self that leads us throughout the task of individuation, thereby taking us through whatever paths are necessary."[22] The designer Joan Nicolau builds this Jungian concept of individuation into his jewelry (Photos 6.9 and 6.10). Nicolau thinks that in our current consumer society, which tries to standardize us and influence our behavior subliminally, jewelry should contribute to revealing the individual and challenge the process of self-identification. Self-discovery and self-interrogation are the main characteristics of quantum jewelry. The motto of jewelry changes from "having" to "being."[23]

Nicolau's jewelry is neither a pretentious accessory nor a symbol of social status. For him, a type of social equality that promotes individuation, and thus is able to cultivate and defend personal differences, is the basis of the new society we should build. And he believes that quantum jewelry will contribute to

Photo 6.9. Joan Nicolau. Broach made with gold, silver, and common stone, 1999. Used with permission.

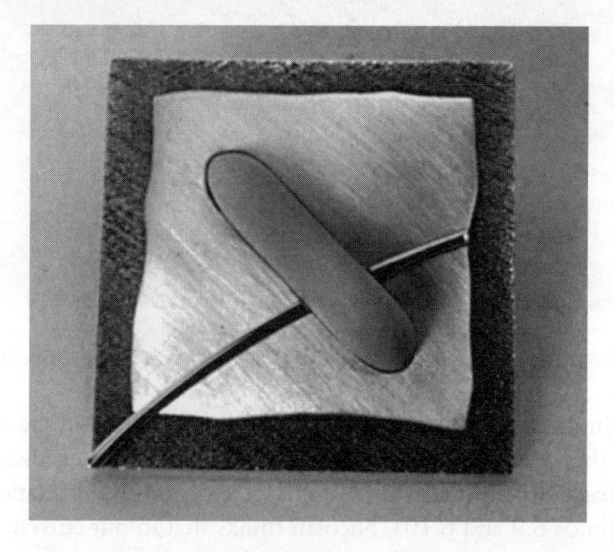

Photo 6.10. Joan Nicolau. Broach made with gold, silver, and common stone, 1999. Used with permission.

the construction of this society. Accordingly, Nicolau establishes a dialogue among the most diverse materials: the noblest elements are mixed with the most humble. In this way, every material is given an ancestral dignity that was lost when humans quit worshiping their surrounding world and became enamoured of objects and things. Stone and wood, for example, are revealed to be precious pieces of jewelry.

Perhaps, the most important revolution that our group advocates pertains to our belief in a reality that until recently was not considered real but now claims this status. Our group is aware of the need to look at the "base of the iceberg"—that which cannot be seen but sustains, shapes, and provides a foundation to what is seen—and anything that is born, rooted on, or developed from this base. Our works are the result of scientific observation and experimentation, as long as this science is understood as liberated from the mistaken task of having to generate theories that outline universal behaviors. From science, however, we adopt its beneficial humanistic thrust and respect for the mysteries of nature. This modus operandi gives way to a type of thinking that is free and open to various types of knowledge, produces ways of living that reconcile individuals with themselves, promotes individuation and the production of personalities that are not attracted easily to fads and fashions, and makes us part of a quantum soup[24] where opposing poles are integrated and we are attached to distant galaxies.

Before ending, we want to make clear that quantum aesthetics is not the only possible path in today's art world. This approach to art is simply the logical consequence of no longer being self-absorbed and becoming interdisciplinary. For example, the boundaries that separate painting, photography, and sculpture are currently being erased, and thus specifying the discipline to which an artwork belongs is becoming an increasingly difficult and inconsequential task. New technologies, images, and sounds, and even the Internet, coexist with two-dimensional works. For example, computers and philosophy, numbers and poetry, and matter, thought, and virtual reality transcend their respective domains and cooperate with one another in both the research processes and the development of their final products. As a result, specialization looses its appeal. In this moment of renovation of the language of art and art criticism, the work of the Quantum Aesthetics Group is born.

NOTES

1. Translated into English by Manuel J. Caro.
2. Leonard Shlain, *Art and Physics: Parallel Visions in Space, Time, and Light* (New York: Morrow, 1991), 427.
3. Quantum Aesthetics Group. Quantum Aesthetics Group's *E-m@ilfesto*. 20 July 2000 ‹http://www.terra.es/personal/lucschok/estetica/emailfestoeng.htm›.
4. Ibid.
5. Ibid.
6. Matila C. Ghyka, *Estética de las Proporciones en la Naturaleza y en las Artes* (Barcelona: Editorial Poseidón, 1983).
7. Ernst Cassirer, *Filosofía de las Formas Simbólicas*, vol. 2 (México: Fondo de Cultura Económico, 1972). English version: Ernst Cassirer, *The Metaphysics of Symbolic Forms: Including the Text of Cassirer's Manuscript on Basic Phenomena* (New Haven, CT: Yale University Press, 1996).
8. Albert Einstein, *Relativity: The Special and General Theory* (New York: P. Smith, 1931).

9. For example, see Albert Kosko, *Fuzzy Thinking: The New Science of Fuzzy Logic* (New York: Hyperion, 1993).

10. Rupert Sheldrake, *Seven Experiments That Could Change the World* (London: Fourth Estate Limited, 1994).

11. Ernst Cassirer, *Filosofía de las Formas Simbólicas*, 9.

12. Ibid., 89.

13. Personal conversation with Andrés Monteagudo, June 2000.

14. David Bohm, *Wholeness and the Implicate Order* (Boston: Routledge and Keagan Paul, 1980), 190.

15. Mircea Eliade, "De lo Sagrado en el Arte Contemporáneo," *Arte y Parte* 25 (2000): 66.

16. Personal conversation with Agustín Ruiz de Almodóvar, June 2000.

17. Quantum Aesthetics Group, *E-m@ilfesto*.

18. Anders Engman, *Mikael Fagerlund* (Sweden: Herausgeber Editor, 1997), 26.

19. David Deutsch, *La Estructura de la Realidad* (Barcelona: Anagrama, 1999), 85. English version: David Deutsch, *The Fabric of Reality: the Science of Parallel Universes—and Its Implications* (New York: Penguin, 1998).

20. Personal conversation with María Caro, April 2000.

21. Jorge Luis Borges, *El Aleph* (Madrid: Alianza, 1971). English version: Jorge Luis Borges, *Collected Fictions* (London: Viking Penguin, 1999).

22. Gregorio Morales, *El Cadáver de Balzac* (Alicante: Epígono 1998), 25 26.

23. Personal conversation with Joan Nicolau, May 2000.

24. For example, see Chung-Liang Huang, *Quantum Soup* (New York: Dutton, 1983).

CHAPTER 7

Individuation and Social Group[1]

Juan Antonio Díaz de Rada

The central ideas of this chapter have already been commented on in the preceding pages of this volume. And the chapters that follow will provide you with a greater knowledge regarding the world of quantum culture. I first came in contact with the world of quantum culture through a book by Gregorio Morales: *El Cadáver de Balzac (Balzac's Corpse)*[2]. Morales's writing made an impact on me, among other reasons, because it coincided exactly with the situation and needs of that time period in my life—a synchronicity, maybe?[3]

With respect to the concept of synchronization, and in a leisure activity, I had the opportunity of posing a discussion that was intended to be a brief experiment to contrast the aforementioned concept with a story.[4]

I had a debate with some of the people who comprised the group: José Gabriel Ceballos ["Specifically I speak of the narrative, which is my field of work. Because it permits us to search for ways of introducing the concept of immateriality into literary creation, which are completely different from those in use up to the present"],[5] Morales ["How can a physical principle be translated into an aesthetic structure?"],[6] Graciela Elizabeth Bergallo

["I only know of my time
The fascination and the terror
The joy and the sadness
Everything
In a never-ending succession
Of contradictions
That is me . . ."][7].

Or Julio César Jiménez ["Without a doubt the creator is whatever he or she wants to be"][8]

My colleagues' responses guided and helped me to continue deepening my thought in these issues. Here, through these pages, and continuing with the experiment, I intend to overturn some ideas regarding the relationship between the *subject and the world*: "Individuation and Social Group."

THIS IS THE STRIFE FOR INDIVIDUATION

"Jung described the process of individuation as the growth and expansion of the personality that is produced when one arrives at and realizes who one is intrinsically."[9]

Or as Horacio Ejilevich Grimaldi explains: "The theory of Jungian individuation seeks to integrate the human Being who has split into **Shadow** and Person, in order to place a new synthesis into motion that responds to what '*one is supposed to do in this world.*' "[10]

"The process of individuation is an ultimate psychic process that requires entirely special conditions if it is to become consciousness. It is rather the beginning of an evolutionary path."[11]

(THE REASON HE WAS BORN!)[12]
WE WILL ADDRESS THE CONCEPT OF SOCIAL GROUP FURTHER AHEAD

In this attempt to explain, it occurs to me from daily, routine experience that... yes, from the experience that comes from your half-awake eyes at seven in the morning. So I was saying, it occurs to me that: We could live a good experience by creating this article shared between us.

In other words, it would be incredible that, just like the process of individuation never ends, this handful of ideas and experiences were to remain open and nourishing indefinitely the creative spirit in every person.[13]

This discussion will apply the following tools of analysis:

➢ To reflect
✓ To verify
❖ To conclude[14]

This said, we will try to mix "art," "life," and "science" in seven keywords, on which I will continue to expound.

KEYWORDS

Light-Shadow,
Myth-Science, Initiate-Err, Stability-Catastrophe, Past-Future,
I-Other, Child-Parent.
Total number of words: 7

Individuation and Social Group 129

> *"The printed word cannot be taken literally" (Serge Leclaire).*

"Little Red Riding Hood goes through the forest. The wolf awaits her"
"Now we are hunting"

I will start at the beginning, in order to describe the end, the way all fairy tales are written.

- ✓ Once upon a time there was: An individual.
- ✓ Once upon a time there was: The Other.
- ✓ Once upon a time there was: A social group.
- ✓ Once upon a time the individual was the social group and the social group the individual.

❖ Individuation is the exercise of liberty. Individuation is self-knowledge, which never ends. Individuation is the process of becoming a person, "an individual"; an individual of enunciation and not of the enunciated. This individual is an end in itself and not a means. ***The individual is the MEANING*** per se. Individuation transcends the paradox "Nature-Culture." Individuation is the incarnation of myth in the person. Individuation, as was defined by Nietzsche and Einstein, transforms the Person into *himself*, that is, Time. Territory is permanently wrench from the unconscious by individuation. Individuation allows persons to articulate *symbolically* the unconscious with the use of the signifier itself. An individuated person feels united "archtypically" to the universe, being *one*, "especially" one self. Through individuation, one reaches the *ultimate origin of information*.

Individuation ***is not*** merely to be a live cadaver, but to be the permanent oscillation between life and death, wisdom and craziness—deconstruction. It is the passageway from the reign of biology to the reign of ideas.

> *"The relationship of ethics to aesthetics is the same as man to the environment, integrating what he is with others, without allowing the essence of either one of them to go up in smoke."* "Individuation is to submerge one self inside of one self in order to blend into the sociocultural environment that envelops us."[15]

KEYWORD NO. 1: LIGHT—SHADOW (IMAGE, SOCIAL I, DUALISM)

Pedro wakes up at seven in the morning like he does every day. The routine comforts him by making him immune, dead-alive, to the passing of time. He turns on the radio: the usual voice enters through his ears informing him of lives and deaths, catastrophes and successes and political figures, all this mixed in the cocktail of publicity. He enters the bathroom ready to perform his daily hygiene. Dulce's death has had him wound up for a month now with latent and desperate anxiety. *He moves toward the mirror, expecting to see the same familiar, spiritless face.* He examines his expression in the mirror and considers, quite anxiously, that the image of his own face is not reflected. He rubs his eyes and looks again. In slow motion the mirror traces the face of someone absolutely unknown to him. Instinctively, he runs his hands all over his body, in order to be certain *that he is the same as always*—the same bony hands, the same thin legs, covered in hair, the same giving and accepting hands full of veins, like the branches of a tree that refuse to succumb. The disgust of life is too big. **Definitely he is still out there.** For an instant, he even thinks that he is not yet awake and is immersed in an impertinent nightmare. He heads to the kitchen, playing along with the situation, as if nothing has happened. But the kitchen window, with its set of lights in a still ongoing night, informs him that "this is serious." His expressionless face is spread between the chimneys of the apartment facing his and leaves a phantasmal empty space in the place that used to be taken by his head. *Without reflection and with a gesture of desperation he squeezes his hands to his face, a mask that continues incomprehensibly in its place.*

- ➤ One of the first steps of individuation is through the looking glass, like Alice. That is, the affirmation of one's self by means of a simultaneous negation-affirmation—the destruction of the "social I."
- ✓ Identity, within the traditional framework, establishes the construction of a "subject" as deeply alienated from himself and the environment. In the traditional conception, the subject is part of the enunciated. However, it is never the subject of the enunciation. Given this portrayal, identity translates into an "speculate-illusory" image. The subject is where the environment asks it to be, but it neither *is* in profundity what it *is*, nor is it capable of maintaining its vital location, its *essence*.
- ➤ It is necessary to contemplate, erasing the illusion of the fictitious image, what pertains to our light, in tension, our shadow. It is a *tense* dialectic where the potential and the actual exist simultaneously in the realization of the so-called reality.
- ❖ The crucial dimension in the comprehension of one's self in Time. To be in Time or be Time represents a crucial difference. In the first case, we are sentenced to experience space; in the second, space is a pathway that does not exist, but rather *becomes* as one lives. In the words of the poet Antonio Machado: "*Traveler, there is no path, / the path is made by walking.*"[16] Among lights and shadows, struggling between them, that lost face exists, that made us change the dichotomous logic for the following expression: "A" and "not-A" exist simultaneously. If the *traditional logic* cannot grasp this, *psychologism* (psiquismo) can. Through the use of integration, metaphors, and dialectics, the paradoxical *reality* is deconstructed.
- ✓ "Know thyself" is not only an intellectual process. This activity places in tension all of a person's shady and luminous qualities. It integrates the intellectual and the emotional, while reaching the central tension. **That** is where the "subject" opens its atemporal *essence*, because the **subject** is an end and not a means.
- ➤ The deconstructive dialectic offers by definition both a tolerant epistêmê, a holographic model for visualizing the world,[17] and a base for respectful relationships. The reason is that the individual is the whole, but also the part. Individualism is the residue of an illusory and estranged ethics-aesthetics that is very far from the language of the gods, that is, myths.

❖ "Who would ever know if Thomas—perhaps one would have to think here of the 'incredulous'—had more faith than the rest, rejecting his own belief and asking to be able to see and to touch? But that which touched his body of meat and bones, was it what he was looking for when he asked for a resurrected presence? Was the illumination that transfigured him not as much shadow as it was light?"[18]

KEYWORD NO. 2: MYTH—SCIENCE (ANAMNESIS, HISTORY, UNICORN)

One time they told him how difficult the maneuver would be to see the image of his thought. But what he never would have imagined is that, without realizing it, he had the ability to manipulate the gradation of his light and shadow, until he could make a part of himself disappear.

"Alright, then . . . no face. So what?"

He has some cookies for breakfast—those that are insipidly gross and help one to maintain a figure—and a extremely watery coffee: "light," also called "American." His stomach tells him something is not right. From the viscera comes out a little grunt in the form of a burp. Bah. What loathing! In his imagination, imagin-ari(a)lly, surges an eagle that plays with the wind while descending to Earth. He becomes completely absorbed in the sequence. *From the bottom of the lavatory a pen(is)etrating voice tells him:* **"We ate from the Apple of Science and not from that of Life."**

I'm feeling so sick because the pizza had gone bad or because I got mad at Elvira for her nauseating rationalism. I believe that my stomach is not feeling so well.

I'm a little anxious—he was saying to himself.[19] I should return to Ángel's tavern and have as good a time as the last day. I had forgotten my lighter; it had sentimental value. I returned to the last place I had been. It was Ángel's tavern. I entered the living room of my house, the day was over, and Ángel was the same as always: serving the house vermouth and roe-deer-sirloin appetizers. *The Minotaur and the Mermaids wait for all of us.*

➢ Individuation is the fission in thought, the run through the forest: The forest where goblins live on your shoulder and whisper what life has to say. The forest is where we celebrate the mythical encounter and we are told how to use our tools. Most of all the pliers, which help us to extract the thorns from ALL of us. "And my soul, my soul watches proudly at the doors of all the dead."[20]

➢ "In the beginning: the death and revolution of societies. If the event occurs at the beginning, when more energy than is available is lost, cultures die. The major part of cultures designated as primitive have been dying at the hands of Power and history; when death closes in, words stop. The death of historical cultures is Revolution: For revolution to be possible, words must be converted into writing."[21]

✓ What is myth but *"poetic science."*[22] We think we know more than we really do, and we know more than we think we do, even though we cannot formalize this knowledge. Do myths tell us less or more about the knowledge that we claim to have in order to make decisions in the world?

- ✓ Petri's networks are probably the best ways of representing decision-making processes in complex systems.[23] Use them!
- ❖ What is science but a *defensive and paranoid* map of reality, a reality to which we have no other access but through interrogation? Only every once in a while, through knowledge—which is something more than just science—we can contemplate all the variables in a sunset. Yes, those variables that escape rationalism and enter, also, through the way of emotion. How do we formulate a kiss of someone that loves you and even interests you?
- ➢ Maybe by using Zadeh's fuzzy logic?[24]

KEYWORD NO. 3: INITIATE-ERR (SENSELESSNESS, MISTAKE, MEANING)

Behind the Madrid of business, money, lack of scruples, and hurriedness, subsists the essence of a town in which the neighbors socialize in kitchens, living rooms, and bars. Places where time is associated with smells, forms, colors, object, and sensations lost in an ancestral infancy. Both Ángel's tavern and the living room of my house are lost, like my lighter. At the end of the wall is history, in the form of keys, ash trays, photographs, army emblems, books from the 1930s, wooden boards for washing clothes in the cold water of a stream, and mirrors that reflect forgotten images: mine, yours, ours, in front of, behind, at the fence. I believe that I have wakened in a day *with no meaning*.

Look at what reminds me now of Ángel's tavern!

But my dream. Yes today's: The chorus of old women has already attended to other beginnings. One woman raises her voice singing: Haim, Haim!, she says. The rest of them sing the chorus at the beat of some drums, and raise their voices and arms and make them tremble; their hands and palms are extended into the fog. *Here I am, preparing to go into the Entire Night.* The trip is humid, dark, and ploughed by multiple traces—radical knots of coarse skin. Only faces illuminated by the eyes of the old women can be seen. It's impossible to orient oneself in this thick space, where the horizons are all the same because of their singularity. A spiral wind wraps around me, originating at my feet and weaving the breeze into a whirlwind. It's impossible to get close to the rest of the people, for they all feel the same endless passage of time. Haim, Haim!, they repeat again and again. Each time the Word sounds, I'm closer to the mountain. My feet barely graze the ground, and the wind carries them through the air. It's like a heard shepherded by the old women; they exhale the Word in procession. Arriving at the mountain, one of the old women reaches for a hatchet that is stuck by its handle—a black trunk with knots—in the ground. The old woman cuts wood, while others put it into a pile. They light the bonfire and the flames start dancing around. The flames are blue, yellow, red. Sometimes they turn greenish and finally they decompose into a pale purple color that inhales the air from the sky. All of a sudden the drums become quiet and the voices stop. A deafening silence spreads out, only to be broken by the spark of death—that is, wood. *Suddenly all of the women become quiet, the leader raises her arms until*

Individuation and Social Group 133

the palms of her hands face each other without touching, making the figure of an hourglass.

- ❖ Truth emerges from the uncertainty that is associated with speech.[25]
- ➢ Finding meaning is crucial: "By finding ourselves, by finding that particular trait that nobody on earth has except us, we can really produce something that is truly our own. Therefore a paradox is produced: the more individuated we are, the more our solidarity with our fellow men and women expands; the greater our compromise with truth, the less our fear of violence and intimidation becomes, and the more our generosity grows. Racism and sexism also disappear, for we understand that the other (man, woman, child, shaman . . .) is a part of us, and thus any attack against him or her is an attack on our own integrity."[26]
- ✓ "In order to give meaning to events, we tell the others: We tell stories. Myths."[27]
  "Let's say that you have resolved the mystery of creation. What is your destiny?
  Let's say that you have been able to strip truth of all its covers. What is your destiny?
  Let's say that you have lived one-hundred happy years
  And may you live one hundred more. What is your destiny?"[28]
- ❖ An initiation is a shamanic event in which the subject finds his or her being through the Other. The subject can be in a wheel running in circles eternally, until he or she achieves the quantum jump of individuation: "to visualize the negativelly entropic universality of his/her being." This process drives the subject, not to be a part of destiny, but to shape destiny.

KEYWORD NO. 4: STABILITY-CATASTROPHE (EQUILIBRIUM, RUPTURE, DISEQUILIBRIUM)

Deep down I like Elvira. She's good-looking, that's for sure, and she makes me nervous. That's also true. I'm going to give her a phone call. However, my diarrhea is no accident, it's due to either the pizza or her. She is a sublime interlude between the neck, waist, and legs. What is certain is that she "turns me on." I need to calm down; that's what Ángel says: "You have to *balance* yourself."

Deep down what I desire is—well:

1. To recover my face but I don't know where to look.
2. Not to torture myself over Dulce's death.
3. To speak to Elvira. I will arrange to meet with her in Ángel's tavern today.

I believe that I'm a little disturbed. I don't know: My work at the university bores me, it's so routine, always the same scientific keys, always the same rigor, always the same articles and the same stories. Couldn't we do something more creative?[29] Couldn't we play at decomposing little by little what so many rational minds have composed;[30] tear the veil from the temple out of my own inner-senselessness, make the meetings to which I am invited tremble with my questions, or jump over the tables of the executives in the offices and slide calmly through the toboggan of life.

What life? The truth is that it's a bit *unbalanced*.

But, that's why my "teddy bear" still interposes itself between me and reality. And maybe the Other. I haven't been able to even reconcile myself to this revolting kid who keeps bothering me as soon as I drop my guard. I like the little kid: Shorts, dirty knees, long bangs, and two eyes that penetrate reality to its core; but the fucker is full of traumas.

I need to recompose my space, yes, and my time. Yes, as Ángel told me: "*Equilibrium in the disequilibrium.*" But, in order to do this, I need to *break* away, I need to create, I need to provoke a permanent catastrophe.[31] Surely I'd need to become a carpenter, painter, electrician, or plumber. A career in which my hands can feel and touch the object of my creation. What do I want?

Perhaps to break from my history, to stop the wheel, to LIVE.[32]

➢ The process of individuation is a permanent effort to assume our intimate essence in relation to a world. From a shamanistic perspective, this process requires that we "fight." If previously the concept of initiation was discussed, now a clear example of part of the method is demonstrated.

✓ In day-to-day experiences, we live a process of permanent adaptation. We can function with routines, like live cadavers, or live well and apprehend our authentic *vocation*, a word unfortunately insulted in the present.

❖ Bachelard says once: "Complexity relaxes me."

❖ See the following figures.

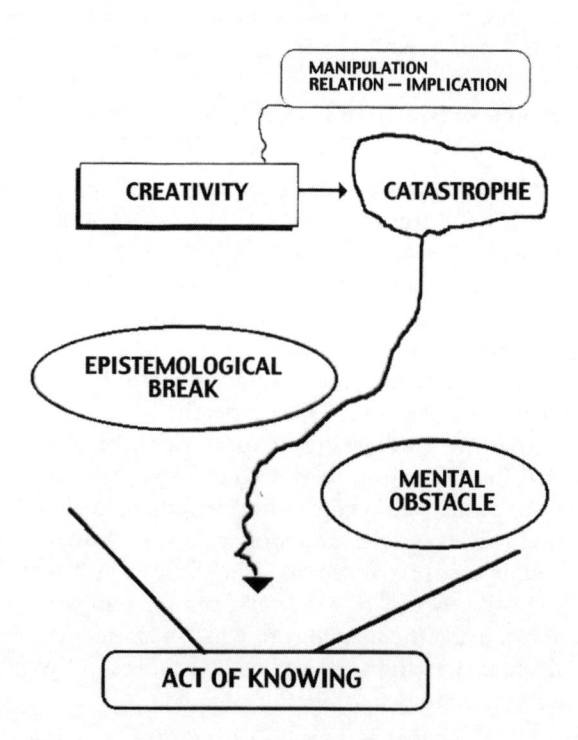

Figure 7.1. Creativity, which involves knowledge, occurs within tension and epistemological breaks.

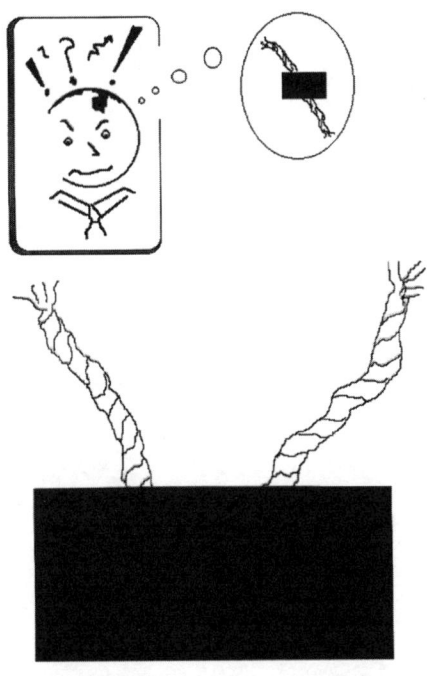

Figure 7.2. We perceive reality in a reductionistic manner.

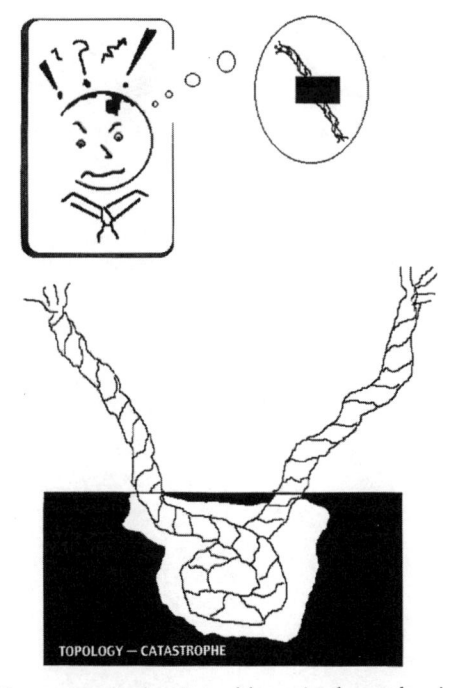

Figure 7.3. We try to see what is problematic about the situation.

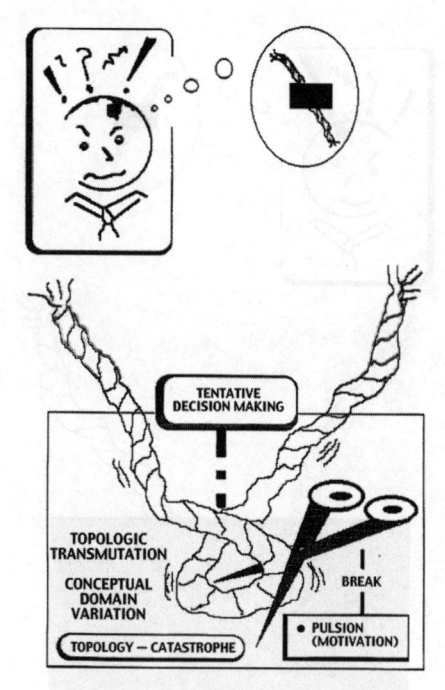

Figure 7.4. I try to bend reality—with little success. This fact prompts me to make a decision.

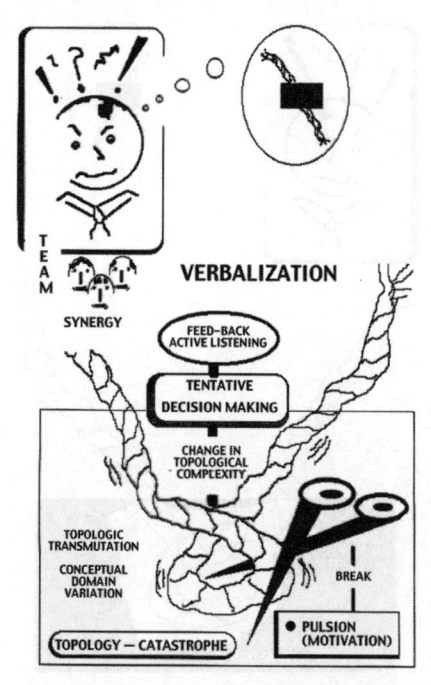

Figure 7.5. From verbalization and synergy between external and internal efforts emerges an outcome that can help us transcend the problem.

Individuation and Social Group

KEYWORD NO. 5: PAST-FUTURE (SYNCHRONICITY)

If it weren't for Pedro dragging himself around for Dulce, always running from himself, *we could enjoy*—He and I—an infinite number of things.

- *(The doorbell rings)* Who's there?
- "The Club."
- What Club is that?
- The Readers' Club, Ma'am.
- I'm not in a readers' club.
- Ah no! Well, according to my list you are.
- Well I'm not: "I am not the person you think I am."
- Well, I have your book.
- Which one?
- *Merengue* . . . Pietro's.
- Excuse me: "I'm not in the club."

Elvira heads toward the dining room thinking about Dulce, about Pedro . . . About the Readers' Club and merengue.

- *(The telephone rings)* Hello?
- Hi, Elvira.
- Hey! Pedro. I was just thinking about you. It's been so long. How are you?
- No kidding, what a coincidence, you were just now thinking about me! How are you?
- Me, wonderful.
- Well, if you have some time tonight I'd like to invite you to dinner at Ángel's tavern.
- Of course. What time?
- 10 o'clock.[33]
- See you later then.
- See you soon.

➢ At the dawn of time words were composed of sounds particular to each species and of five vowels, which the Goddess emanated through the pores of her skin, on the earth just like in heaven. And it occurred at that time that humans began walking across the earth and opened their eyes to the heavens, discovering the splendid face of the Goddess. The moon sent the wind, for the first time, in order to refresh the forehead of humans. The wind began to touch, with soft hands, the forehead, eyes, and lips of each and every one of the creatures. In unison, all of the beings yelled the name of Astar. In all of history, a greater thunder cannot be remembered. Mother sent the rain in order to suffocate the burning passion that came from their throats, but it was already too late: the fire was already created and started ascending to the sky. The sun that was hidden, as it always was, began to emerge from between the mountains. Where the sun was born humans stuck an enormous rock, on which was carved the first circle of history. Days and nights began to succeed one another, and the present decomposed into past and future, but only for THEM. This is how they tell the story about the rising of the sun. This is how they chronicle the birth of the calendar. Since then, human beings have wandered between the past and the future, only sometimes have they been able to grasp the present, and only then have they been able to understand the authentic language of creation.

➢ "We're animals: Birds fly and we speak."[34]

KEYWORD NO. 6: I-OTHER (IDENTITY)

Yours, mine, the other's	The other as necessary other.	The other to be human.
The other in desire	The other at the other side.	The other because it's necessary.
The other in necessity.	The other on the shore.	The other, the other, the other YOU.
The other in a compassionate clash.	The other behind the surge of waves.	
The other in the sweetness of a kiss.	The other to look at you.	The other tells me and the other, that life is ME.
The other in not being able.	The other is Other with you.	
The other in vital breath.	The other is me, and without him, I don't exist	
The other in breathing.	The other to reflect you.	

I'm told: That the sunset is a delicate circular square of colors.
I'm told: That the surface of the moon is the color blue with the smell of jasmine.
I'm told: That I'm a bolt with defects, but I can enter into any nut.
I'm told: That the taste of grass is green—thick.
I'm told that the path of the soul has no other limit but I, HIM, HER.

I'm told a story; one of those that let you sleep in peace.
One of those that open your chest to drop in two grams of hope.
One of those that inundate your eyes with white tears.
One of those that make you feel near an abyss of nostalgia:
Of the maternal womb.
Of warm rain on the lips.
Of birth with the pain of death.
Of those stories,
Of those singular, unique stories, born for You.

And the Other is You without wanting,
without knowing, you tell the story again, and you make numbers, counting the days, about the Time that life has given you.
And . . . in the end you account for yourself by telling your story.

Individuation and Social Group

> "Story, story, María Sarmiento.
> She went to shit and was taken by the wind."[35]
> Wind/Time/Story. Other time, You, I.
> **. . . And three 3 is born: Creation**

- Ángel! Who wrote the poem on the door?
- Hey Pedro! Good evening. Your friend Elvira wrote the poem.
- I'm meeting her now for dinner, at ten.
- What should I prepare for you two?
- Whatever you might have. For now, give me a vermouth with this wonderful cheese you have here.

Facing him, a mirror returns his reflection again. An image a little transfigured. The decision to see Elvira has interrupted his obsessive circle of sadness for Dulce.

➢ Here are some historical notes on the concept of "identity" supplied by Morales in a personal conversation. I subscribe to them:

- In primitive towns, identity is erased by the group and the individual lives in the collective unconscious of the clan.
- In Western societies after the Renaissance, the individual is identified by common features that are shared with other individuals, such as nationality, profession, sex, and not by characteristics that differentiate them.
- The scientific societies during the nineteenth and part of the twentieth centuries appropriated the concept of Truth, dividing the world into Real and False, Legitimate and Internal (dualism), privileging the white race above the rest, and among white people, men over women, and among men, scientists over humanists.
- In postmodern societies, the elements of cohesion have disappeared completely and a ferocious and egotistic individualism is exalted within a negligent hedonism.

KEYWORD NO. 7: CHILD-PARENT (INDIVIDUAL)

Ten o'clock on the dot. Elvira enters Ángel's tavern, with such grace and a splendid smile! Pedro watches her absorbingly. Perhaps he didn't remember her as so beautiful, or had never looked at her with desire, with those sparkling eyes. Chills run up and down his spine. The encounter with his shadow, sadness, loneliness, child, face had its compensations.

- Elvira!
- Hi Pedro.
- You look . . . !
- How are you?
- Better now.

They sit down to enjoy the meal that Ángel prepared for them. Pedro can't take his eyes off Elvira; he starts to note a certain twitching in his eyes; and some physiological reactions occur with minimal modesty. A chill runs up the back of his neck with such a force that it makes him bend over the table.

- What's wrong? Are you okay?
- Yes, it's nothing, lately I've been feeling a little strange.
- Well, you were undressing me with your eyes.
- I'm sorry . . . you already know that so much solitude isn't good for anyone.
- No, don't be sorry, you don't know how you've made me feel, man.
- Hey! . . . The norms are the norms, one can't walk around eating women with their eyes. That's what I think.
- It depends on how one "looks."[36]

Elvira smiles trying to calm him. Deep down, and on the surface too, she likes him. Pedro's response to her smile is a caring gesture. They continue with the dinner, without saying a word, looking at each other. Ángel prepares dessert. Life CONTINUES.

> ➤ The beginning of pleasure versus the beginning of reality. The discovery of the child, his/her impulse, recreation versus the norm (the Super-Ego). Norms versus anarchy. And something deeper: the resolution of guilt based on adult responsibility and freedom. That is what the process of individuation does for us: Integrating our child and our parent, responding before the world without guilt, but rather with responsibility—"You've made your bed, now you have to lie on it." There is no better synonym of arrogance than to feel guilty. Our current fragmented society empties our language of content, fills us with euphemisms, and generates guilt in us. Then, in a paternalistic way it pardons us so that we depend effectively more and more on the mechanisms of alienation. Castration and guilt are fundamental concepts that impede the subject from being integrated into social groups and acting freely, while truncating the discursive mechanisms of human beings. These ideas make us the same; they homogenize us. They don't take into account the specific character of each being. We homogenize our own differences. In this we are the same! "An archipelago is a set of islands that are united by that which separates them."

> ✓ It would be interesting as a reflective and active experience, if you, my friend and reader, were to realize a metasubjective assignment. That is to say, evaluate your conduct throughout the day, from the perspective of responsibility, not as an exercise of confession. Surely you will be surprised at the amount of things that you've done without knowing why. Contemplate your authentic desires, concerns, passions, decisions, and relations with the world, and accordingly you will have initiated the journey to individuation. However, reflection without action, or action without reflection, serves little purpose. Act according to your reflections. The most important thing in the process is that your words and your actions are joined.

> ❖ In the *process* of individuation one begins to produce a true *language*,[37] one that is personal but also in harmony with all of creation. In the manifesto of the Quantum

Aesthetics Group, some relevant reflections are posed that can help us to advance our own system of "expression."[38]

Dear Reader:
We have arrived at the end of our seven keywords.
Life continues encountering us at every second. And we encounter life.

"In order to live on a war footing . . . Seconds, out . . . Seconds, out."[39]
"In order to Live."[40]
" For freedom, I bleed, I fight, I survive."[41]

As I commented at the beginning of this handful of reflections: This chapter does not intend to be more than an experiment to mix in the crucible of *our relationship*. **"Art, life and science."**
I will start at the end, in order to describe the beginning, the way all fairy tales are written.

✓ Once upon a time the individual was a social group and the social group the individual.[42]
✓ Once upon a time there was: A social group.
✓ Once upon a time there was: The Other.
✓ Once upon a time there was: An individual.[43]

NOTES

1. Translated into English by Leeann Hunter. [Because of copyright issues, the Spanish originals of the poems discussed in this chapter have been omitted. Only Hunter's translations of these excerpts have been included.]

2. Gregorio Morales, *El Cadáver de Balzac* (Alicante: Epígono, 1998).

3. Carl G. Jung, *Synchronicity: An Acausal Connecting Principle* (London: ARK Paperbacks, 1985).

4. The bells begin to ring, slowly, obscurely. It's 9 o'clock in the morning. It's a Monday in autumn. They toll for others, but also for me. My name is Pedro. I never would have imagined that it would be so extraordinarily strange crossing the threshold of the dead. No, it's not that I'm dead, not at all, at least in part; I'm simply shaken up by Dulce's funeral. Dulce was my friend and her loss causes the sound of the bells tolling to provoke a sour taste in my throat. I've barely slept two hours. I find myself grieving. Bong! It's a special kind of grief. Bong! The ringing resounds in my head. Bong! The sound makes me dizzy. Bong! I turn around, my consciousness throbbing; I point myself toward the village. The village is round like the kind of sun kids draw. I enter Antonio's tavern, which looks a little like Angela from my neighborhood. On the radio I hear: "Yolanda . . . Yolanda . . . Eternamente Yolanda" ("Yolanda . . . Yolanda . . . Eternally, Yolanda"). I've listened to this song many times with Dulce, but this time the notes are amplified in my head. I feel like the voice of "Milanés" slides in through my ear reaching down until it knots itself inside of my stomach. Nausea and more nausea. Always nausea every time I try to challenge reality. Dulce has simply died. I'll never see her again; now I can never speak with her, nor

take walks with her, nor will I be able to see her almond eyes. I have a cup of coffee and decide to go down to Madrid. I think I'm going to spend the day in the park; I can't stand so much absence. I take leave of Antonio, without speaking, with a gesture of my eyebrows. Antonio looks at me with consoling eyes, like those of the moon on my nightly walks. I start the car. Automatically, the cassette tape starts playing and Pablo returns to singing the same tunes: "Yolanda . . . Yolanda . . . Eternamente Yolanda." Yes, now I remember: I left the tape on yesterday when I came from Madrid for Dulce's funeral. With my eyes on the road, and my thoughts streaming through my memories, I make my way toward my peaceful park. My park! It always has a new corner in which to talk, rest, and enjoy. I cross one of the side streets that leads to Plaza Elíptica. A six-year-old child runs right into my path. I swerve to avoid hitting her. With my nerves rattled I step out of the car, look at the little girl, and fortunately she's okay. She lifts herself up and looks at me with her almond eyes. She takes off running, turning back to me with a smile. I lean against the hood of the car; my arms are trembling. My memories take me back to my childhood and an image of a warm young girl appears, with dark hair and almond eyes. She always joined me when we played our childhood games. Her name was Yolanda and one afternoon she lost her life, right by me, after banging her head up against the playground equipment. YOLANDA was so SWEET! I enter the car; I'm almost to the park. I can't avoid letting out a smile at remembering how this little girl had disappeared running, so "lively," behind the corner. Stepping out of the vehicle and contemplating the sky, the clouds outline the face of a woman. I notice the presence of my two friends more closely than ever and this comforts me.

5. José Gabriel Ceballos, e-mail of the Quantum Aesthetics Group, Thursday, June 24, 1999.

6. Gregorio Morales, e-mail of the Quantum Aesthetics Group, Sunday, June 27, 1999.

7. Graciella Elizabeth Bergallo, e-mail of the Quantum Aesthetics Group, Sunday, June 27, 1999. Bergallo cites here one of her own poems.

8. Julio Cesar Jiménez, e-mail of the Quantum Aesthetics Group, Tuesday, September 7, 1999.

9. Alfred M. Freedman, Harold I. Kaplan, and Benjamin J. Sadock, *Compendio de Psiquiatría*, 2nd ed. (Barcelona: Salvat, 1987), 153.

10. Horacio Ejilevich Grimaldi, *Los Orígenes de la Psicología Analítica*. 31 December 2000 ‹http://www.psiconet.org/jung/psi-analitica.htm›.

11. Carl G. Jung, *Arquetipos e Inconsciente Colectivo* (Buenos Aires: Paidós, 1970), 170. English version: Carl G. Jung, *The Archetypes and the Collective Unconscious* (London: Routledge, 1991).

12. ("Pá lo que ha nacío ¡Vaya!") Camarón de la Isla and Tomatito, *Camarón con Tomatito. Paris 1987* (MERCU, 1999 [CD]).

13. The reader is welcome to participate in the debate and controversy surrounding quantum aesthetics by sending his/her contribution to the following e-mail address: esteticacuantica@teleline.es

14. "A scientific theory tends to constitute a closed system. It's a paranoid formation. The redundancy of science is in the theoretical component (storage device), while the practical, empirical or technical component opens itself to events. Accordingly, the dialectic of interaction between both components assures the spiral motion

of scientific progress." (Jesús Ibañez, *Del Algoritmo al Sujeto: Perspectivas de la Investigación Social* (Madrid: Siglo Veintiuno, 1985), 17).

15. Elías Escribano, personal conversation, December 1999.

16. See the original Spanish version in Antonio Machado, "Proverbios y Cantares," in *Poesía y Prosa*, vol. 2, *Campos de Castilla (Madrid: Espasa Calpe/Fundación Antonio Machado), 575*. English version: *Antonio Machado, The Castilian Camp* (Isle of Skye, Scotland: Aquila/Phaethon, 1982).

17. Jeffrey Mishlove, Arthur Bloch, and Karl H. Pribram, *The Holographic Brain* (Oakland, CA: Thinking Allowed Productions, 1988 [videocassette]).

18. Michel Foucault, *El Pensamiento del Afuera* (Valencia: Pre-Textos, 1997), 38. English version: Michel Foucault, *Maurice Blanchot, the Thought from Outside* (New York: Cambridge, MA: Zone, 1987).

19. "Anxiety is a feeling of fuzzy fear, unpleasant, many times wandering, that is accompanied by one or more recurrent corporeal sensations. It is a warning sign that advises us of immediate danger and permits us to confront the challenge." Freedman, Kaplan, and Sadock, *Compendio de Psiquiatría, 357.*

20. Saint-John Perse, *Anábasis* (Paris: Gallimard, 1926). English version: Saint-John Perse, *Anabasis: A Poem* (New York: Harcourt Brace Jovanovich, 1977).

21. Jesús Ibáñez, *Más Allá de la Sociología. El Grupo de Discusión: Técnica y Crítica* (Madrid: Siglo Veintiuno de España, 1979), 157.

22. "Allegory is a paraphrase with a conscious content. A symbol, on the other hand, is the best expression possible for a unconscious content of which we have a premonition but do not yet know." Jung, *Arquetipos e Inconsciente Colectivo*, 12.

23. For example, see Grzegorz Rozenberg, ed., *Advances in Petri Nets, 1993.* (New York: Springer, 1993).

24. L. A. Zadeh, "Fuzzy Sets." *Information and Control* 8 (1965): 338–353.

25. Jacques Lacan, *Écrits* (New York: Norton, 1977), 305–307.

26. Gregorio Morales, e-mail of the Quantum Aesthetics Group, June 1999.

27. Jesús Ibañez, *Del Algoritmo al Sujeto: Perspectivas de la Investigación Social* (Madrid: Siglo Veintiuno, 1985), 100.

28. Omar Khayyam, *Rubaiyat* (Barcelona: Plaza and Janes, 1969). English version: Omar Khayyam, *Khayyám: A Critical Edition* (Charlottesville: University Press of Virginia, 1997).

29. A good education necessarily breaks away from organized education. See Paul K. Feyerabend, *Farewell to Reason* (New York: Verso, 1987).

30. We should unlearn the majority of what we have learned, and learn what is usually not taught. See R. D. Laing, *The Politics of Experience* (New York: Ballantine, 1978), 131–135.

31. "Nevertheless, it's possible that certain universal principles will be discovered that are applicable to different fields, even if they adopt a specific and unique form in each field. 'Order through fluctuation' (1980) by Prigogine and the theory of catastrophes (1975) by René Thom constitute two important examples. Keeping in mind these reserves, now we can discuss the relationship among diverse investigations of consciousness and the holonomic focus of the universe and the mind." Stanislav Grof, *Psicología Transpersonal: Nacimiento, Muerte y Trascendencia en Psicoterapia* (Barcelona: Kairós, 1988), 110.)

32. Morales, *El Cadáver de Balzac*, 36–52.

33. "Only in morphogenesis is there diachrony in the strong sense of the word (time is irreversibly oriented . . . of morphogenesis, evolution, or revolution. . . . Learning, as a process of personal or political liberation, enhances the negative entropy irreversibly; we go from one structure to another that is logically superior)." Ibáñez, *Del Algoritmo al Sujeto*, 15; see also the discussion on morphogenic fields in Chapter 1 of this volume).

34. J. M. Díaz Charcán, personal conversation, January 2000.

35. ("Cuento, cuento, María Sarmiento. / Se fue a cagar y se la llevó el viento.") This is a popular Spanish story that is sometimes told to kids as the last bedtime story.

36. ("Depende de cómo se 'mire.' "). The author is playing with the verb "mirar" (too look, to see) in the two senses of looking: (1) looking at something physically and (2) thinking about something in a certain way.

37. "Nothing is more true than the following: Everything is in a process of continuous evolution and change, even truth itself." Joachim Schickel, *Gran Muralla, Gran Método: Acercamiento a China* (México, D.F.: Siglo Veintiuno Editores, 1972), 255.

38. Quantum Aesthetics Group, *Quantum Aesthetics Group's E-m@ilfesto*, 20 July 2000 ‹http://teleline.terra.es/personal/lucschok/estetica/emailfestoeng.htm›.

39. Para vivir en pie de guerra . . . Segundos fuera . . . Segundos fuera.") Luis Eduardo Aute "Segundos Fuera," in *Segundos Fuera* (Ariola, 1989 [CD])

40. ("Para Vivir.) For example, see Pablo Milanés et al., "Para Vivir," in *Querido Pablo* (Ariola/Eurodisc, 1985 [DC]).

41. (Para la libertad sangro, lucho, pervivo.") See Miguel Hernández, "El Herido, I," in *El Hombre Acecha* (Madrid: Cátedra, 1985).

42. "Descriptively speaking, a psychological group is defined as one that is psychologically significant for the members, to which they relate themselves subjectively for social comparison, and acquisition of norms and values (i.e., with which they compare to evaluate themselves, their abilities, performances, opinions, etc., and from which they take their rules, standards and beliefs about appropriate conduct and attitudes), that they privately accept membership in, and which influences their attitudes and behavior. In the usual terminology, it is a (positive) *reference* group and not merely a *membership* group as is defined by outsiders, i.e., it is not simply a group which one is objectively *in*, but one which is subjectively important in determining one's actions.

This definition is not intended to restrict our awareness of the full reality of the human group—of the fact that collective life is biological, sociological, political, ideological, historical, and so on, not just psychological—but it does indicate a deliberate limitation in subject matter and purpose." John C. Turner, *Rediscovering the Social Group: A Self-Categorization Theory* (New York: Basil Blackwell, 1987), 1–2.

43. "[B]oth biological and social systems are autopoietic [H. Maturana and F. Varela, *Autopoiesis and Cognition*, 1979; F. Varela, *Principles of Biological Autonomy*, 1979]—are constructed by themselves; their only product is themselves. . . . In quantum physics—Microphysics—the subject who measures and the disposition of the measure modifies the object measured (uncertainty)," Jesús Ibáñez, *El Regreso del Sujeto: La Investigación Social de Segundo Orden* (Madrid: Siglo XXI, 1994).

Ibáñez is referring in this excerpt to: Humberto R. Maturana, and Francisco J. Varela, *Autopoiesis and Cognition: The Realization of the Living* (Boston: D. Reidel, 1980); and Francisco J. Varela, *Principles of Biological Autonomy* (New York: North Holland, 1979).

CHAPTER 8

The Interpretative Foundations of Culture: Quantum Aesthetics and Anthropology[1]

Graciela Elizabeth Bergallo

> A poetic awareness of the world precedes, as should be the case, a rational assessment of objects. The world is beautiful before it is true. The world is admired before being examined.
> —Gaston Bachelard[2]

INTRODUCTION

Traditionally, various sources of absolute knowledge have been advanced to explain reality that eventually seeps into every aspect of existence. In the end, this base becomes a fictional source of security in a nebulous universe that lacks secure meaning and is gradually institutionalized as culture and a hegemonic source of power. From this general foundation emerges positive science, which provides a totalized rendition of knowledge that discredits human intuition. This style of science is predicated on a belief in empirical and objective facts that can be obtained only if persons abandon a side of themselves that makes them human—their feelings, sense of purpose, and values. Therefore, human beings are subjected to a science that dehumanizes and converts them into slaves of a style of knowledge that is detached and impersonal. However, the production of other forms of knowledge is stigmatized. Any alternative view of reality is converted into something crazy or perverse by the custodians of the "official definitions" of normalcy. In other words, alternative forms of knowledge are destroyed because they are presumed to constitute the dark side of life that threatens the clear and objective picture that is traditionally presented by science.

From its inception, anthropology has showcased types of knowledge and ways of understanding reality that science has dismissed as primitive, simply because they are different from European conceptions of reality. Often, these cultures are based on worldviews that maintain an intimate connection with nature, the stars, or spirits, and thus these outlooks clash with the rational version of the universe embraced by Westerners. Regularly, these alternative ways of constructing reality are explained as typical responses to either a lack[3] or an excess[4] of environmental structures. Furthermore, shamanism and other spiritual elements are classified similarly as "religions of the oppressed" or as "new age" beliefs, both of which reflect the ethnocentric and mechanistic paradigm that dominates the West.

A different approach to anthropology, one based on the validation of alternative knowledge bases and the discoveries of quantum physics—as opposed to positive science—would be better able to provide a more culturally sensitive understanding of these religions' phenomena, and, at the same time, illustrate that the old views held by anthropologists are erroneous. Specifically, Western science is only one of the many ways of thinking about the world, and represents only one method of organizing and interpreting facts in order to explain human existence. In a certain sense, positive science represents simply another culture.[5] The ambiguity of the universe at the quantum level demonstrates that reality is the result of choices made by humans and, thus, can be modified by further action. Simply put, according to quantum physics, reality *does not exist* in itself but *is made*. As a result, persons are not anchored to "progress" or "universality," as is presumed by Western science. That is, although positive science has made possible and continues to reinforce the idea that physical (and social) reality is uniform and deterministic, the principles of quantum physics have placed this image of existence in doubt.

In fact, by paying attention to the tenets of quantum physics, we begin to notice that the outlook of positive science is founded on certain mechanisms and formulas such as dualism. Furthermore, these assumptions conceal the subtle differences among physical and cultural phenomena, suppress emotions, and stifle the ability of humans to perceive, think, create meaning, and dream. Thanks to science, the Western world has been systematically deanimated.[6] This destiny will befall all societies once they know the truth, or more accurately, once they are given the truth. However, is nothing true about the lives that people may desire and accept as valid? Is the religious dimension of existence,[7] for example, unrelated to the ability of humans to decide what is significant and experience this reality within the subjective, idiosyncratic, and (ir)rational lives that they may decide to live? That is, the division of life into two separate domains, with religion, myth, and intuition on one side and reason and science on the other, is only a trap.[8] Supporters of quantum aesthetics argue that this opposition is unwarranted: their claim is that humans are so closely involved in creating reality that the latter could not exist without the former, and thus reason and intuition include each other. According to this nondualistic world, reality is no longer identified simply with matter, and thus imagination, symbolism, and myth are understood to be indispensable parts of human existence. Additionally, the possibility exists that the unfathomable, the unexplainable, and the mysterious—as Michael Leris calls it, "the rare capacity of transforming the worst deserts into playgrounds"—can be realized.[9]

Each culture is the product of a hierarchy of values and hidden elements—a particular imagery—that forms a tradition of knowledge that is linked to a unique history. This history may come from the experience of the colonizer or the colonized, from a constant struggle with the environment or a friendly relationship with it, or from rapid social change or lasting social stability. Whatever the case may be, these special ways of constructing reality and responding to challenges cannot be explained in all of their complexity by the rationality imposed by positive science, because this approach is insensitive to how persons define themselves and their respective realities. Social factors such as art, myth, or religion, for example, cannot be studied properly within the strictures imposed by dualism, and thus cannot be readily examined by positive science. In the words of Immanuel Kant, these phenomena are part of the numinous realm, and therefore cannot be reduced to something that is simply material. And due to their elusive character, these elements of social existence have been stigmatized time and again by science.

To a certain extent, anthropologists have incorporated this insight into their work for many years. What is innovative about quantum anthropology is that it does more that merely recognize the legitimacy of various cultures, but additionally reveals how contact with the West has resulted in many of these societies adopting a colonial mode of constructing reality. Not only have these non-Western cultures been abused, oppressed, and often destroyed, in many cases they have lost their ability to enter different layers or reality through visions or dreams, or view reality as consisting of overlapping and dynamic fields, as is proposed by quantum theorists. As a result of adopting the myopic space-time world imposed by the West, the existential depth of these societies has been lost.[10] Following the censure of alternative forms of knowledge, positive science becomes the undisputed source of reality. As a result, Western societies are left without many legitimate alternatives, while the diversity represented by other cultures continues to be attacked and reduced. Quantum aesthetics represents another of these alternatives. This theoretical approach, moreover, not only strives to diversify the reality of the West, but also to protect other cultures that have been labeled regularly as primitive by Western scientists.

The aim of this chapter is to apply certain principles of quantum aesthetics to the anthropological study of a neoshamanistic movement found among the Toba, which is an ethnic group that resides in the Chaco region of Argentina.[11] This exposition will help readers to understand how certain phenomena, such as dancing, the wind, and ecstasy, connect the Toba to the implicate or hidden order of the universe.[12]

A TOBA MOVEMENT:[13] DANCE AND RESISTANCE

As a reaction to colonization, the Toba, Pilagá, Mocoví, and Mataco in northeast Argentina have produced a syncretistic movement that includes several traces of traditional shamanism. This activity is characterized by the appropriation and reinterpretation, with respect to their traditional worldviews, of key elements of the dominant culture and nation-state, specifically Pentecostalism. In Latin America, both millennialist and messianic movements have constituted a frequent, albeit not exclusive, response to colonial expansion.

The violent process of colonization in the Chaco, during the end of the nineteenth and beginning of the twentieth centuries, required that the indigenous people modify their traditional way of life. The exploitation of sugar, tannin, and cotton placed this region under the economic control of foreign powers, thereby initiating social, ecological, and cultural changes that modified dramatically the history of those who lived in this area.

One of these changes took the form of a neoshamanistic movement around the middle of the 1950s.[14] The ceremonies that involved dancing, ecstacy, and dreams were not suspended during the periods of cutting, harvesting, and processing of the sugar cane, despite how the time cycles were altered in these societies by colonization. Additionally, some of the youth in these movements became involved in the active struggle to recover and replenish the land. That is, the neoshamanistic movement that still exists nowadays, became involved in the attempt to resist Western interference in these societies. Popular spirituality and religious language became a potent mixture in the "silent rebellion" against colonialism.[15]

Similar to today, dancing was very important and had multiple meanings in the ancient rites. Many aspects of traditional Toba ritual have been synthesized in present-day cults: dancing, healing, collective expressions of values, communal interpretations of dreams, common responses to the problems of subsistence, and amusement. In certain situations, some aspects of these cults, such as colors, were given new meanings in order to avoid persecution and repression. As a result, certain symbols had opposite meaning for those inside and outside of the group.

In order to obtain *harmony, knowledge,* and *power* a person had to dance. It was maintained that the "message," with respect to knowledge and power, came to the Toba directly without the mediation of the colonizers. Likewise, the voice of the *spirit* is understood to be announced through the wind, clouds, and birds. The most controversial element, and the most authentic trait of these indigenous people, is the dance, which constitutes a direct mode of access to visions. Shamanistic knowledge revolves around these elements, and is directly related to the construction of reality that occurs in these groups.

The Toba's outlook came under attack by Cartesian dualism, which is at the root of Western hegemony and requires that the world be divided into two irreducible substances, known as mind and body.[16] However, in the midst of the conflicts provoked by colonization, the Toba succeeded in maintaining and reaffirming their own vision of the cosmos that places the mind and the body on a single continuum. In this regard, they resisted the worldview that the colonists tried to impose. Indeed, the dance involves the transcendence of dualism and represents a relationship between the material and nonmaterial dimensions of existence. Alfred Métraux indicates that the *fiestas* and other ceremonies, in which dancing occurs, have been extremely important in the revivalist movements in Latin America.[17] As Estela Ocampo explains, "the point of unification between nature and the beyond is the *fiesta*, which embodies a synthesis of all expressive forms."[18]

The *pi'oxonaq* (Toba shamen) were central figures in the ancient rites where dancing occurred.[19] They were guides and custodians of their communities and were responsible for maintaining the harmony of their respective groups. Furthermore, these men opened the doors of the numinous realm through ecstasy, but they did not act alone. All of their activities followed the life cycle of the community, which was understood to be a living landscape on the nature-human-spiritual continuum. The community, in this sense, was not reduced to the economic categories that the colonists tried to elevate in importance: nature is part of this community and is not something simply to be exploited.[20] In this sense, dancing takes place in connection with the most important events of a person's life, the annual cycle of a community, and the cycle of nature to which humans have adjusted.

The colored ribbons that are utilized in rites are considered to be signs. These signs, moreover, originate in dreams or visions. Diverse meanings are attributed to these colors that relate, for example, to the cosmological, gnoseological, and erotic dimensions of existence, in addition to representing codes of communication. Colors illustrate better than many other modes of expression the polarizing potential of symbols; in a color the normative and socially sensitive sides of symbolism are united. Through the presence of emotion and value a symbol gains meaning, while at the same time this style of expression acquires an affective tone. A color is symbolic in a manner similar to a hologram or hieroglyph. Colors are powerful and radiate energy that affects persons in both positive and negative ways.[21] Despite Newtonian reductionism, in several cultures colors have emotional, visual, and symbolic connotations.[22] Colors reveal the extraordinary power of concentration, have unquestionable importance in human cognition, and mediate the relationship between persons, on the one hand, and the world and individuals, on the other.

IMAGES OF TIME AND THE COSMOS

> We are not of a different species than the sand that we step on or the clouds that hover over our heads are. We are all cut from the same fabric, and in all parts, on any corner of this cloth, the fabric intervenes in the dance. Prigogine states this as follows: "Matter is not inert. It is alive and active."
> —Michael Talbot[23]

According to José Imbelloni's interpretation,[24] there exists a similarity between the Toba's outlook and that of Andean cultures with regard to the construction of time. These societies point to four ages that are connected to four cardinal points, and their chronological union fills the great void of the passage of time. The fifth age, correlated in a spatial sense with the conception of the center, is that which people are presently living. Each of the three planes of the spatial universe are divided into four sectors that are united by the world's mast or central pole. Each of these sectors represents a cardinal point whose value is determined by a season, an element, a cardinal animal, a virtue, an activity, or a god. To give another example, the Zuñi maintain this same tetravalent organization, thereby associating each of the cardinal points with an element (water, earth, air, or fire), a season of the year, and a color (yellow, blue, red and white.) This conception of time and space coincides in part with some of the Toba's testimonies. The colors blue, white, yellow, and red, which represent sectors of the sky or distinct planes of reality, appear in other Toba myths and rituals.[25] Imbelloni argues that the cosmogonical cycle of Andean cultures is an example of a type of intuition of the world that is common to all protohistoric civilizations.[26]

The Toba's concept of time is cyclical. Birth is followed by death, and death is followed by a rebirth. Future and past represent a mythical time. "Looking to the future" is impossible because the future could already have past. The spirits of the dead can return to us in different ways; plants die in winter to be reborn in spring. That is, time does not follow a straight line from past to future, as it does in our Western and Newtonian world. Time is not ordered: yesterday, today, and tomorrow are mixed and succeed one another in no specific sequence.

The traditional universe of the Toba consisted of five essential "strata" that are inhabited by three main categories of beings: celestial beings (*Piguem'lec*), terrestrial beings (*'Alhual'ec*), and aquatic beings (*Ne'etaxaal'ec*).[27] However, subsequent to the introduction of various elements of Christian mythology, ambiguity was introduced with respect to the nature and identity of the powerful entities, even within the Toba groups. In the Toba's universe, the earth is found at the center. Beneath the earth lies another level, referred to as "the other earth," where there exist beings who live according to a rhythm opposite to that of humans, although the landscape is quite similar.[28] The night is the moment when the dwellers of "the other earth" awake and wander around the world of humans. The West is also called the "sky," because this is where the earth and the sky meet. Beyond and above this level exists another sky where the stars and other celestial beings are located.[29] The belief that the universe consists of a series of superimposed compartments is espoused by many indigenous groups in Chaco. Métraux also describes two Mataco myths that portray the world in terms of three connected strata, where similar persons reside who periodically exchange their positions.[30]

The conception of space expressed in these rituals breaks from Newtonian renditions of space that depict this concept as uniform and infinite. Moreover, every segment of this space is divided into various parts that have determinate positions within the totality of the cosmos. This spatial totality resembles a box of well-defined and linear forms. None of these spatial parts, in other words, can intersect with any other because they are all located at different points. This rendition of space, accordingly, follows the Aristotelian principle that "A" is the same as "A," but different from "B," for both are located at different places.

Nonetheless, the understanding of space exposed by the Toba rituals referred to earlier is more consistent with the concept of space espoused in quantum physics. In this theory, space is divided into regions with limited and flexible dimensions that can intersect with little difficulty. Space is not homogenous, but rather consists of multiple planes and dimensions that are interchangeable. Because this rendition of space is antidualistic, space is complex and polyvalent. In this sense, space does not separate elements but rather unites them and thus instigates an exchange between various fields. The quantum perspective does not limit the number of spatial domains, but rather makes them interchangeable, imbues them with subjectivity, and allows them to change according to how persons construct their spaciotemporal lives. Earth, "the other earth," and the sky, for example, are spaces that do not exist in isolation from one another but rather are interconnected.

That is, despite its many variants, these images of the cosmos and of temporal seasons are not far removed from some of the hypotheses advanced in quantum theory. David Bohm argues that subatomic particles are correlated in a way that defies traditional conceptions of the world, thus suggesting that a level of reality is present that lies beyond the quantum. Although subatomic particles are perceived as being separate, in reality they are merely extensions of the same fundamental "something." These particles participate in a constant "dance of interacting parts."[31] All things in the universe, accordingly, are interconnected. Stated otherwise, in what Bohm calls the explicate order, things might seem separate and unrelated, but in the implicate order this separation vanishes as all things constitute a reality bereft of discontinuities.[32] The universe should no longer be understood to be a machine, but rather as a multidimensional hologram, in which the past, present, and future are folded and exist simultaneously. Three-dimensional space and time should be viewed as projections of a more profound order, one where concepts such as location lose consistency.[33] In Toba rituals, space and time are united in such a way that the West, yesterday, the past, the "there," the future, and the self are united in an existentially complex and multidimensional reality that is supposed by the nondualistic perspective of quantum aesthetics. The following testimony of an indigenous person offers an illustration of this reality.

If I get ill, if I am grave, the family situation disequilibrates; and not only does the family situation disequilibrate but the general situation of the entire community as well. There is a feeling of malaise in every community member when this takes place. A family feud produces a disequilibrium in various other aspects of reality. We live amid these sorts of circumstances.

The world, accordingly, is conceived as something that is round, in which all culture, nature, and humankind resides. Because humanity belongs to nature and resides within nature, humanity needs nature. If something goes astray, everything disequilibrates. The *pi'oxonaq* is in charge of restoring the equilibrium of things.[34]

This indigenous testimony reflects the *connection* and *inseparability* of all phenomena in the universe—where any pertubation in one sphere reverberates in others: a drought or an illness, for example, produces a state of generalized disequilibrium. According to this outlook, the borders between the individual, the social (or intersubjective), the natural world, and the sacred are diluted. The privileged space of action taken by the shaman to reach the necessary equilibrium and harmony in the life of a community is a collective ritual.

Ken Dychtwald affirms that every aspect of the universe is expressed vibrationally and contains the knowledge of everything that exists within the cosmos.[35] Because vibrational expression is also a manifestation of pure information, every particular facet of the universe has the capacity to be recognized in every other aspect within the overarching hologram. Every part or aspect does not exist as an individualized affirmation, but rather also contains a rich warehouse of information, that is, a basic understanding of the existential nature of the rest of the universe. This holographic principle is central to quantum aesthetics.

THE COSMIC TREE AND BEINGS OF AIR AND WIND

> The poets of fire, water, and earth do not produce the same kind of inspiration as does the poet of air.
>
> —Gaston Bachelard[36]

The Toba share with other groups from Chaco the belief in a gigantic tree that connects the sky with the earth.[37] Nonetheless, due to diverse circumstances, communication with the celestial plane is reserved for shamen.[38] According to traditional beliefs, the *pi'oxonaq* works under the guidance of a *Itaxayaxaua*, which literally means the "companion who speaks." The shaman can undertake a "flight" toward the other strata of reality with the help of his spirit companions. The idea that the power of the *pi'oxonaq* depends on his *ltaxayaxaua* is deeply rooted in Toba culture.[39]

Circular dancing around trees was maintained constantly in the testimonies of Chaco tribes in ancient times. The tree symbolized a fusion among the different cosmic planes. This is consistent with what Mircea Elíade calls symbolism of the center. Within the macrocosmic plane, this communication is represented by an axis (tree, pillar, or staircase), and within the microcosmic plane by any locale or space that expands through different levels by projecting itself toward the "center."[40] Much like the symbolism of a tree, dancing generates a center that makes this communication possible. The cosmic tree expresses the world's sacred character, fecundity, and enduring nature. The tree is related to the ideas of fertility and initiation, in addition to the absolute reality and immortality to which dance is also connected. Only the shaman, however, transforms this cosmic conception into a concrete experience through a *flight* in the strictest sense of the word. Shamen can establish a real communication among the three cosmic zones. Here, once again we witness how the Newtonian conception of space is trespassed, superseded, and replaced with a more dynamic and flexible version. This space is not encapsulated within a rigid box, but rather constitutes an open path to our subjectivity and embodies the world that we create.

According to Elíade, the dreams, visions, hallucinations, and myths that have as their central theme a *flight* or ascension and that belong to the generality of humankind cannot be portrayed adequately through psychological or historical explanations.[41] And perhaps these considerations that cannot be explained reveal the true situation of man in the cosmos? Indeed, the flight represents the most dream-related of metaphors. Air, precisely because it is impalpable, invisible, and uncatchable, offers a high degree of expressive liberty to those who wish to use it symbolically. In fact, because it appears to be without limits, the symbolic use of air is much more liberating than other, more concrete and immediate referents such as earth and water.[42]

Unlike all other elements, air seems unseizeable, and this condition, if only conceptual, makes air fascinating and mysterious. Gaston Bachelard maintained that physical elements are the usual ways in which matter is rendered in both dreams and fantasies.[43] In Toba cosmology, the wind could not be defined—and neither could any other element—in an exclusively natural way, for in this culture the world is defined as unified/unifying and continuous; in other words, the separation between the natural and supernatural is untenable. The wind and its manifestations provide a passage for spirits; souls can thus wander around during dreams.

The four winds represent spatial orientations and temporal seasons. The idea is that history is written in the wind, including the past, present, and future:

The wind is where our history is written. Our memory lies in the wind. In the four angles, cardinal points, is where our secret resides. It is here where the lives of our ancestors are found. But one must discover, open his/her mind, and allow spirituality to penetrate oneself. The wind is what transmits and keeps our secrets. It (the wind) is like a code; sometimes it consists of words. There are spirits of the wind. There was once a dance for the wind.[44]

The wind propagates things and effluviums; it contains memories, suggests that everything is fused and inseparable, and promotes the continuity of the cosmos. The wind conveys the idea that disequilibrium in one sphere can have impact on others. Stated otherwise, the air connects everything that exists and supplies a holographic perspective of reality. The air contains everything that has existed throughout history. The air is thus the memory of all humanity—what Rupert Sheldrake, consistent with quantum theory, calls morphogenetic fields.[45] Here again, we see the close connection between quantum theory and some of the Toba's ideas that have been considered primitive by prior ethnocentric renditions of science.

BODIES, SOULS, AND SPIRITS

> The hypothesis of [the existence of] the spirit is no more fantastic than that of [the existence of] matter.
>
> —Carl Jung[46]

> If the *what* of consciousness consists of pure information, in a universe of meanings, the *where* of consciousness is alternatively converted into *nowhere* and *everywhere*. In a universe that is infinitely interconnected, consciousness can be where ever it wants to be: in the *dances* of the neurons in our brains, . . . etc.
>
> —Michael Talbot[47]

Dance is a symbol for other symbols: dance evokes a fusion of cosmic planes that allows the operation of internal and external realities, and perhaps even changes both of them. The analogic-poetic fluctuation not only transcends borders, spaces, and time, but also links the real and the imaginary through metaphors and the concrete and abstract through homologies and isomorphisms. In this way, novel forms of organizing knowledge and thought are fostered. In all acts of invocation, the imitation of a person or spirit (his/her movements, voice, or physical appearance) allows us to reach their soul, or essence. Mimesis has the power to evoke and can be considered one of the key characteristics of Toba dance, for these dances have the analogic-holographic character described by Edgard Morin.[48]

Lpaqal alludes to *the shadow* that is associated with "the other life." The *Lpaqal* appear at night and can take on the appearance of whirlwinds. Certainly, the spirits of those leaders who have influenced the past and have been kind and powerful continue to serve as spiritual guides in the present and offer knowledge and power. In general, people make reference to them by affirming that they are still alive and continue to influence the lives of the Toba. In this manner, the barriers between life and death, past and present are diluted. Similarly, quantum physics makes similar claims about temporal connections in the Western world. As Talbot affirms,

> three dimensional time and space are not, according to Bohm, the only processes that can be better understood if they are viewed as folded and unfolded in the interior and exterior of the implied order . . . [I]n the superhologram of the universe, the past, present, and future are tightly folded and exist simultaneously. This suggests the possibility that,

at some point, the most distant past can be revealed in the superhologram, and even retrieved.[49]

Within the holographic paradigm, every moment or every aspect of time exists always in any place. In this manner, time constitutes a dimension (or many dimensions) that is full and alive. Every moment coexists in a cognizable and holographic relationship with every other moment.[50]

In the Toba world, during the ecstatic trance (*ntonaxac*), a person's soul (*lki'i*) can be separated from the body to investigate facts, provide insight into the causes of a disease, or search for lost souls. This process also refers to the continuous or natural-spiritual permeability where such changes are possible. *Lki'i* means "image." This image is not different from the person; it is a "copy" of the person that is separated from the body during dreams or visions. The soul is not understood as being distinct or separate from the body. Morin points out that non-Western renditions of the self are universally united to the experience of its "double," and thus adhere to the principle of identity/distinction with regard to all forms of knowledge.[51]

These ideas and testimonies express the existence of souls or doubles, spirits and transformations, and lead to the image of a person who is explained in terms of the ways in which individuals understand their relationship to nature and the social group. This concept deviates sharply from that of individuality—as defined by Louis Dumond—that is so dear to the Western world.[52] The individual is neither unitary nor universal. The person should be understood more in terms of the concrete actualization of symbolic conceptions about humans and their place in the universe.

ECSTASY, VISION, POWER: THE MAGIC IN DANCE

> The advocates of easy thought usually say that people use magic because they are not capable of explaining certain phenomena. From this point of view, how could anybody be seduced by a discipline whose only objective is saving us the trouble of searching for the causes of thunder?
> —Alejandro Dolina[53]

Within Toba culture, the term "power" (*napinshic* or *jawan*) loosely refers to political processes as we understand them nowadays, but also, and more importantly, to human relationships with the universe that affect control over the forces that preserve health and harmony. Power does not derive from a being or an entity, but rather is constituted as a result of the structural ordering of key elements such as partner-spirits, powerful objects, and cosmic levels.[54] Dance, dreams, and ecstasy—and their symbolic contexts—generate the necessary conditions for such power to be revealed. The notion of synchronicity that Carl G. Jung discussed can help us to understand this power.[55] Synchronicities refer to the significant and parallel co-occurrence of psychological states with one or several physical events. Dreams and visions are considered to be existential dimensions that are ontologically comparable to daily life, and thus are produced within a context that is understood to be "real." Power is the force that is capable of producing or counteracting disorder. The shaman's role is to mediate between the community and this power, so that equilibrium and harmony are reinstated.

Both the mimetic character of magical events and the role of colors as referents are based on the idea that the signifier (the part) can contain what is symbolized (the whole) holographically. These elements belong to a system of thought that is symbolic and mythological. Due to the chasm produced by Cartesian dualism between imagination and matter, synchronic events have been characterized erroneously as belief, superstition, or something abnormal. However, consciously or not, we suffer the effects of metaphors in our own bodies; we are aware of the meanings of discipline and punishment.[56] These processes and their consequences can be approached from a quantum paradigm because this framework is antidualistic. As Pedro Gómez García explains, "as long as they have objective effects on people and societies . . . the supernatural and magical are parts, in a broad sense, of nature."[57]

The magical character of a dance is thus based on the symbol's ability to evoke and, to a certain extent, contain holographically what is symbolized. For this reason, the magical action over things and beings is exerted through symbols. Magic is based on both the mythological existence of doubles and spirits, and the analogical character of the anthropo-socio-cosmic paradigm. Mimesis is used in hunting, war, and fertility rites. Magic protects societies and regenerates anthropo-socio-cosmic relationships. Accordingly, Morin asserts that mimetic and sacrificial rites have contributed to the production of the great cosmic cycles throughout the world.[58] In sum, human action can influence nature in a noncausal way through what Jung called synchronicities.[59]

STATES OF CONSCIOUSNESS, WAYS OF KNOWING, REALITIES

> The correct way of thinking about the underlying thread is to understand it as though it was *primarily* . . . a dance of intertwined parts and only then as if it was constrained by several types of limits.
>
> —Gregory Bateson[60]

Ecstasy is the most characteristic element of shamanism.[61] The study of this phenomenon defies all of the classical systems of interpretation and explanation. Research into this facet of life requires the transcendence of traditional disciplines, because shamanism exposes an integrative dimension of the human being. Understood this way, the study of shamanism articulates a double understanding of the human—as both a cognitive process and a part of a social system—that challenges the traditional identity of the individual.[62]

Some of the changes the Toba have undergone can be explained by viewing the intersubjective, natural, and sacred dimensions as inseparable. That is, the frontiers between some individuals and others, body and mind, sleep and vigil, imagination and reality, the natural and the sacred, and matter and spirit are permeable. Accordingly, actions, words, objects, movements, and the body itself can evoke and produce metonymic effects, for they are not considered to be distinct from what they represent. There exists a holographic association between the symbol and what is symbolized, the signifier and the signified, and the part and the whole. In other words, for the Toba the symbol is the thing itself. However, the mythological paradigm of integrated duality does not undermine human uniqueness. Identity is now understood to contain holographically the world in which the individual lives. Within the analogic texture of the universe, humans can be inhabited by the forces of nature, while nature, matter, or ideas can be imbued with human subjectivity.

It is usually said that "mythical" and "Western" thoughts are different and distinct, as though they are only alternative ways of conceptualizing any situation. However, as Morin writes, there is no space or time that is exclusively rational or mythical; rather, there exists a duality in the unity of space and time. Those who can use both types of knowledge experience a multifaceted reality as completely normal.[63] Myths cover the huge cracks opened up by rational interrogation and allow the human to be integrated into the cosmos. This existential movement prevents knowledge from being separated from what is known and encourages the ecstatic integration of persons with the essence of reality. That is, we posses marginal cognitive skills that allow us to short-circuit the spaciotemporal contradictions of the phenomenical world. There may exist a reality that is beyond what is knowable, and ecstasy may be our way to enter this region.

As is the case with the Toba, there are other societies that lack frontiers between "the mental state of vigil" and the "mental state of sleep." It is common to hear among the Toba that "this [experience or event] is real because it came through a dream or a vision."[64] In Western societies, however, the unconscious is usually blocked by the "secondary process," that is, consciousness or culture. The latter is structured around a syntactic language that is too restrictive to express the primal contents of the psyche.[65]

Throughout the twentieth century, the basis of a materialism that has reigned since the 1600s and reduced the spirit to immateriality began to crumble at the subatomic level; where there used to be certainty and substance, there is now an enigma. As a result, the mind has begun to be considered an antenna that captures the "transmaterial" messages that are transmitted through a psychological and informational *field*.[66]

Toba's testimonies speak of messages, information, and knowledge that are received. Gregory Bateson, for example, asserts that a system of messages controls self-organization.[67] This is the same phenomenon that Sheldrake tries to explain with his concept of morphogenic fields.[68] Visions and messages come from the unconscious and are imprinted in art, religion, and other cultural manifestations. This mediation of images is essential in order to contact the alterworld inhabited by gods, spirits, or forces of nature. That is, images are crucial in order to create a *fuzzy zone* that integrates the cultural and natural aspects of the human being.

A vision (*l.loxoc*) can only be interpreted by the person who receives it. This person is the only one who can interpret what is written in the wind. This cannot be discovered scientifically, because it is something that one is *given*. This spiritual knowledge is what needs to be used when it comes to the realm of the physical. There are things that happen just as one sees them. . . . It's like blood, we cannot explain why it works. There is a spirit that gives you wisdom, knowledge.[69]

Jung's idea of synchronicity,[70] which has been used regularly by quantum aesthetics, can help us to understand that these modes of knowledge and power are acausal processes that are related to certain psychological conditions or significant coincidences between the psyche and external events. Synchronicities, in this way, may be related to intuition and magic. That is, space and time—the categories involved in causal relationships—can sometimes amount to zero, and thus causality disappears.[71] This acausality that is hidden behind synchronicities could be considered as important as causality used to be. As was the case with causality, the physical sciences—as opposed to the humanities—will probably be the first ones to prove and accept this principle.

Western categories such as "religion" or "economy" are not capable of capturing the way in which Toba's reality is organized. Josep M. Fericgla asserts that the gap between the importance that some people give to their rites and experiences and the lack of attention that anthropologists have given to them is due to the ethnocentric and reductionist attitude present in the usual models of anthropological analysis.[72]

We have seen that, in the cult, there exists a symbolic mechanism that, through dance, promotes ecstasy whereby the body, soul, and cosmos become intertwined. Understanding this process in such a way is anathema to the "identity of the self" that Western cultures deem so important.[73] Within the ritual, the individual travels as a passenger from dance to trance. The characteristics of the liminal (from Latin *limen*: threshold) subject are ambiguous. Victor W. Turner asserts that liminal entities are in none of the usual positions assigned by social conventions.[74] Symbols express a liminal situation. During the liminal state, the *communitas* or communion—intersubjectivity, nature, and the sacred—is expressed. After experiencing this communion, individuals return to a revitalized world.

Achieving a liminal state is usually associated with ritual powers. In this situation of uncertainty, people produce myths, symbols, philosophic systems, and artworks that help them redefine periodically both reality and the relationships of humans with society, nature, and culture. How can we accept that individuals are capable of going outside of their consciousness without understanding this phenomenon to be the manifestation of a precivilized or pathological state or as a means of reinforcing the status quo?[75] As proponents of quantum aesthetics affirm, the idea of a machinelike body is an outgrowth of the metonymies created by a unique mode of knowledge and culture. Specifically, the alleged separation of imagination and reality is another of the consequences of Cartesian dualism. After consciousness began to dominate emotions and the unconscious, the civilizing process left a mark on the body that can be seen in everyday life, that is, in the control of spontaneous impulses.

CONCLUSION

In sum, many movements similar to the one we have been discussing have been considered pre- or pseudopolitical when they are approached from a positivist position that understands politics as a rational and secular phenomenon. However, determining how rational such a process is can only be done by analyzing the belief systems, mythical traditions, and social organizations typical of the people we are studying. That is, we should use methods that are culturally sensitive to the ways in which the people we are examining understand and construct their world. In other words, a critique of eurocentric categories needs to be undertaken urgently.[76] Viewing religion, the economy, or politics as separate and distinct domains of reality is not possible in situations where these regions are experienced as an integrated whole. Quantum aesthetics is based on this realization, and thus its theoretical framework is capable of understanding and accounting for such integration.

The relationship between culture and the quantum paradigm brings epistemology to the center of inquiry, an epistemology that needs to be open or a transepistemology, namely an epistemology within which all modes of knowledge develop in connection with one another without imposition or marginalization.[77] Therefore, anthropology needs to focus on the conditions under which knowledge and culture are produced. Contrary to understanding the self as unitary and autonomous, the complex multiplicity of human beings, the historicity of their consciousness, the integration of the inside and outside of the individual, and the anthropo-socio-cosmic understanding of reality should be recognized.[78]

In this chapter, I have tried to illustrate the relationship that is present between certain quantum principles and the ideas about knowledge and the cosmos advanced by a social movement with shamanistic traces, which continues to exist in the Argentine Chaco. My intention is to show how quantum principles can help us appreciate various cultural phenomena, such as dance, rituals, or myths that used to be denigrated because they are considered primitive, pre-rational, and wild. After being approached from a quantum perspective, the Toba's ability to connect with the implicate order of reality is revindicated. In this sense, quantum aesthetics is trying to recover a skill that the Western world has not only lost, but has also tried to eradicate from the face of the earth.

NOTES

1. Translated by Luigi Esposito, John W. Murphy, and Manuel J. Caro.
2. Cited in José Antonio González Alcantud and Carmelo Lisón Tolosana, "El Aire: Entre Metáforas, Energéticas y Riesgos," in *El Aire: Mitos, Ritos y Realidades*, José Antonio González Alcantud and Carmelo Lisón Tolosana, ed. (Granada: Diputación Provincial de Granada–Anthropos Editorial, 1999), 120.
3. Mary Douglas, *Natural Symbols: Explorations in Cosmology* (London: Routledge, 1996).
4. Ioan Lewis, *Ecstatic Religion* (Harmondsworth: Penguin, 1971).
5. Thomas S. Kuhn, *The Structure of Scientific Revolutions* (Chicago: University of Chicago Press, 1962).
6. In this context, the "term anima" does not carry religious connotations. The point is that science has robbed reality of the human element that gives it life.
7. Although the notion of "the religious" is a central part of the Western tradition, at this juncture this term is not being used in this narrow sense.
8. Jacques Derrida, "Las dos Fuentes de la 'Religión' en los Límites de la Razón," in *La Religión*, Jacques Derrida and Gianni Vattimo, ed. (Buenos Aires: Ediciones de la Flor, 1997), 13. English version: Jacques Derrida, "Faith and Knowledge: The Two Sources of 'Religion' at the Limits of Reason Alone," in *Religion*, Jacques Derrida and Gianni Vattimo, ed. (Stanford, CA: Stanford University Press, 1998).
9. Cited in D. W. Winnicott, *Realidad y Juego* (Barcelona: Gedisa, 1986), 1.
10. See Chapter 4 in this volume.
11. In this chapter, reference is made to aspects of a wider investigation where I analyze more extensively other dimensions of the rites and related issues.
12. David Bohm, *Wholeness and the Implicate Order* (London: Routledge and Kegan Paul, 1980).
13. The Toba, who call themselves Qom o Nam Com (the people), have been the dominant group among the Guaycurú, which includes the Pilagá and the Mocoví, in addition to the Abipone, Mbayá, and Payaguá who are now extinct. After contact with the Europeans, the Mataco-Mataguayo and the Guaycurú are the linguistic groups in the Chaco who have been most influential politically and in terms of their mere numbers. The majority of the Toba are currently found in the central-eastern region of the Chaco, while the Mataco and the Mocoví reside in the western and southern zones respectively. The Toba lived originally in bands, whose names still exist, that comprised various extended families whose reputations were based fundamentally on the use of knowledge and power for the good of the group. See Elmer S. Miller, *Los Tobas Argentinos: Armonía y Disonancia en una Sociedad* (México: Siglo XXI, 1979), 25–27, 53–54. English version: Elmer S. Miller, *Harmony and Dissonance in Argentine Toba Society* (New Haven, CT: Human Relations Area Files, 1980).
14. In 1967, Elmer S. Miller collected testimonies from independent cultists who had not sought support from an official church. During these interviews, the existence of important Toba imagery was discovered that emphasized that dancing and ecstatic trances were central to the cult. See Elmer S. Miller, *Pentacostalism among the Argintine Toba* (Ph.D. diss. University of Pittsburgh, 1967), 127–128. See also, Miller, *Los Tobas Argentinos*, 142–144.

15. Eduardo Rosenzvaig, *Etnias y Árboles: Historia del Universo Ecológico Gran Chaco* (Havana: Editorial Casa de las Américas, 1996), 467.

16. René Descartes, *Treatise of Man* (Cambridge, MA: Harvard University Press, 1972).

17. "The mystical value of the dance is demonstrated in the almost unhealthy significance that it has acquired for the revivalist movements among the Indians in South America. During the Sixteenth Century, and at a level equal to today, the Guarani resorted to frenetic dancing in order to obtain land that is not plagued by evil." See Alfred Métraux, "Religion and Shamanism," in *Handbook of South American Indians*, Julian Haynes Stewart, ed., vol. 5 (Washington, DC: Bureau of American Ethnology, 1949), 559–599.

18. Estela Ocampo, *Apolo y la Máscara: La Estetica Occidental Frente a las Practicas Artisticas de otras Culturas* (Barcelona: Icaria, 1984), 10.

19. The *'oiquiaxaic* constituted a special class of the *pi'oxonaq*, in that they exhibited exceptional knowledge and power through their association with the spirits of the dead or *nnatc* (see Miller, *Los Tobas Argentinos*, 31). They can predict the future and transform themselves into any element of nature or culture because they are omnipotent. The *'oiquiaxaic* are especially able to lead the community during periods of crisis, due to their ability to predict the future. See P. Wright, "Dream, Shamanism, and Power among the Toba of Formosa," in *Portals of Power: Shamanism in South America*, E. Jean Matteson Langdon and Gerhard Baer, ed. (Albuquerque: University of New Mexico Press, 1992), 149–172.)

20. Rosenzvaig, *Etnias y Árboles*, 11.

21. Johannes Itten, *The Elements of Color* (New York: Van Nostrand Reinhold, 1970), 12–13.

22. Xaverio, the quantum artist, is conscious of this power. See Chapter 6 in this volume.

23. Michael Talbot, *Mas Allá de la Teoría Cuántica: Las Polémicas y Audaces Experiencias que Desafian la Teoría Cuántica. Desdibujando los Límites entre Física y Metafísica* (Barcelona: Gedisa, 1995), 149–150. English version: Michael Talbot, *Beyond the Quantum* (New York: Bantam, 1988).

24. José Imbelloni, "Religiones de América n° 10: Las Edades del Mundo; Sinopsis Crítica de Ciclografía Americana," *Boletín de la Academia Argentina de Letras* 11, no. 41 (1943): 159.

25. Alfred Métraux, *Myths of the Toba and Pilagá Indians of the Gran Chaco* (New York: Kraus Reprint, 1969), 25.

26. Imbelloni, "Religiones de América n° 10," 159.

27. Elmer S. Miller, "Simbolismo, Conceptos de Poder y Cambio Cultural de los Tobas del Chaco Argentino," *Procesos de Articulación Social*, ed. E. Hermitte and L. Bartolomé (Buenos Aires: Amorrortu Editores, 1977).

28. E. Courdeu, "Aproximación al Horizonte Mítico de los Toba," *Runa*, 12, nos. 1–2, (1969–1970): 67–176.

29. Miller, *Pentecostalism among the Argentine Toba*. Also in Miller, "Simbolismo,"

30. Métraux, *Myths of the Toba and Pilagá Indians of the Gran Chaco*, pp. 24–25.

31. Gregory Bateson, *Espíritu y Naturaleza* (Buenos Aires: Amorrortu, 1982). English version: Gregory Bateson, *Mind and Nature: A Necessary Unity* (London: Fontana, 1985).

32. Bohm, *Wholeness and the Implicate Order*.

33. Talbot, *Más Allá de la Teoría Cuántica*, 55–60. Similar to Talbot, E. Morin affirms that "we have arrived at . . . a basic microphysical reality where the material is simultaneously immaterial, the continuous is discontinuous, the separable is inseparable, the distinct is indistinct, where things and causes are embroiled together . . . ; the object is no longer found at any specific point in space; the time to come vanishes within it [microphysical reality]. . . . Moreover, everywhere during the course of this century, the corrosion of universality and temporal and spatial absolutism has occurred; these factors have integrated with one another on a macrocosmic scale for which Einsteinean relativity has shown that there exists no universal time independent of observers. . . . In every instance, it seems as though physical reality entails something that is meta- or infra-spatial, meta- or infra-temporal. As it has commonly been assumed, it is plausible that we reside in a polydimensional universe in which we can only perceive three of those dimensions. It is possible that the depths of the physical universe ignore separations, time, space, although these have been and undoubtedly will continue to be the general source of the *physis* of time, space, and separation—hence our idea of the Cosmos." E. Morin, *El Método I: La Naturaleza de la Naturaleza* (Madrid, Ediciones Cátedra, 1980), 76–78.

34. Toba assertions are in L. D. Heredia, "Aportes para la Comprensión del Chamanismo Toba," *Antropos* 90 (1995), 481.

35. Ken Dychtwald, "Comentarios a la Teoría Holográfica," in *El Paradigma Holográfico*, Ken Wilber, ed. (Buenos Aires: Troquel, 1992), 151. English version: Ken Wilber, ed., *The Holographic Paradigm and Other Paradoxes: Exploring the Leading Edge of Science* (Boston: Shambhala, 1985).

36. ("Le poète du feu, celui de l'leau et de la terre ne transmettent pas la même inspiration que le poête de l'air.") In González Alcantud, and Lisón Tolosana, "El aire," 120.

37. Alfred Métraux, *Etnografía del Chaco* (Asunción: Editorial El Lector, 1996), 241. English version: Alfred Métraux, *Ethnography of the Chaco* (Washington, DC, U.S. Government Printing Office, 1987).

38. Nowadays, however, the private character of the relationship between the shaman and his *ltaxayaxaua* has been modified, and everyone now has access to shamanistic knowledge and power.

39. Miller, "Simbolismo."

40. Mircea Eliade, *El Chamanismo y las Técnicas Arcaicas del Éxtasis* (Mexico City: Fondo de Cultura Económica), 212–217. English version: Mircea Eliade, *Shamanism: Archaic Techniques of Ecstasy* (Princeton, NJ: Princeton University Press, 1974).

41. Ibid., 12.

42. A. Lupo, "Aire, Viento, Espíritu: Reflexiones a Partir del Pensamiento Nahua," *El Aire: Mitos, Ritos y Realidades*, ed. González Alcantud and Lisón Tolosana, 258.

43. González Alcantud and Lisón Tolosana, eds., *El aire: Mitos, Ritos y Realidades* (Granada: Diputación Provincial de Granada–Anthropos Editorial, 1999).

44. Toba leader's assertion (as recorded by the author), October 10, 1998.

45. Rupert Sheldrake, *Seven Experiments That Could Change the World* (London: Fourth Estate Limited, 1994).

46. Cited in Edgar Morin, *El Método I: El Conocimiento del Conocimiento* (Barcelona: Anthropos, 1994), 78.

47. Talbot, *Beyond the Quantum*.

48. Edgar Morin, *El Método III: El Conocimiento del Conocimiento* (Madrid: Ediciones Cátedra, 1994), 155–170.
49. Talbot, *Beyond the Quantum.*
50. Dychtwald, "Comentarios a la Teoría Holográfica," 152.
51. Morin, *El Método III.*
52. Louis Dumont. *Essays on Individualism: Modern Ideology in Anthropological Perspective* (Chicago: University of Chicago Press, 1986).
53. Alejandro Dolina, *Crónicas del Ángel Gris* (Buenos Aires: Editorial Colihue, 1996), 145–146.
54. Miller, "Simbolismo,"
55. Carl G. Jung and Wolfgang Pauli, *The Interpretation of Nature and the Psyche—Synchronicity: An Acausal Connecting Principle* (New York: Pantheon, 1955). The concept of synchronicity that is currently used to explain certain phenomena that coincide in the same field without having a common cause has its origin in Paul Kammerer's essay "Das Gesetz der Serie," published in 1919. In this essay, the author defines the concept of "seriality" as the co-occurrence in time and space of events that are meaningfully albeit not causally related.
56. Michel Foucault, *Discipline and Punish* (New York: Vintage, 1995).
57. Pedro Gomez García, "Teorías Étnicas y Etnológicas Sobre la Terapéutica Popular," in *Creer y Curar: La Medicina Popular* ed. José Antonio González Alcantud, and S. Rodríguez Becerra (Granada: Diputación Provincial de Granada, 1996), 245.
58. Morin, *El Método III, 179–180.*
59. Carl G. Jung and Wolfgang Pauli, *The Interpretation of Nature and the Psyche.*
60. In Talbot, *Más Allá de la Teoría Cuántica*, 124.
61. Eliade, *Shamanism.*
62. Josep M. Fericgla, *El Chamanismo a Revisión* (Quito: Ediciones Abya-Yala, 1998), 17–18.
63. Morin, *El Método III.*
64. This is a recurrent expression among the Toba in the region as recorded by the author, April 10, 1998.
65. Josep M. Fericgla, *El Sistema Dinámico de la Cultura y los Diversos Estados de la Mente Humana* (Barcelona: Anthropos, 1989), 19.
66. According to Morin, the overcoming of the separation between matter, body, and brain on the one hand, and spirit and soul on the other, that is, the chasm between the substantialness of being and the immateriality of knowledge, occurs at different levels. Dualism is advanced, first, at the level that is represented by Claude Shannon's concept of information (see Morin, *El Método I*, 340 ff.). According to Morin, information is material for it depends on energy, but it is also immaterial for it is not reducible to matter. Second, dualism is rejected at the level of microphysics. At this level, energy is not substantial and matter (mass) is only one of its aspects. However, the materiality of a particle is only one of its multiple aspects. Third, at the systemic or organizational level, dualism has become inappropriate for the organization of material systems is itself immaterial, nondimensional, and irreducible to mass or energy. And fourth, at the biological level, dualism has been eclipsed because any organizing act includes a cognitive dimension: "At a certain level, both life and mental organizations are the same thing." Morin, *El Método III*, 86–87.
67. Bateson, Gregory, *Espíritu y Naturaleza.*
68. Sheldrake, *Seven Experiments That Could Change the World.*

69. Conversation with an indigenous Toba leader as recorded by the author, April 10, 1998.

70. Jung, *La Interpretación de la Naturaleza y la Psique*.

71. Morin, *El Método III*, 233–239.

72. Fericgla, *El Sistema Dinámico de la Cultura*, 11.

73. Marcel Mauss, "Una Categoría del Espíritu Humano: La Noción de Persona," in *Sociología y Antropología* (Madrid: Tecnos, 1991), 309–333. English version: Marcel Mauss, *Sociology and Psychology: Essays* (Boston: Routledge and Kegan Paul, 1979).

74. Victor W. Turner, *El Proceso Ritual* (Madrid: Taurus, 1988), 101–134. English version: Victor W. Turner, *The Ritual Process: Structure and Anti-Structure* (New York: Aldine de Gruyter, 1995).

75. Márcio Goldman, "A Construção Ritual da Pessoa: A Possessão no Candomblé," in *Candomblé: Desvendando Identidades (Novos Escritos Sobre a Religião dos Orixás)*, org. C. E. Marcondes de Moura (São Paulo: EMW Editores, 1987), 88–89.

76. Alicia Barabas, *Utopías Indias: Movimientos Sociorreligiosos en México* (México: Grijalbo, 1989).

77. Gananath Obeyesekere, "Comment," *Current Anthropology* 38, no. 2 (1997): 271–272.

78. Edgar Morin, "La relación Antropo-bio-cósmica," in *Encyclopédie Philosophique Universelle* vol. 1, ed. A. Jacob (París: Presses Universitaires de France, 1989). Translation into Spanish by José Luis Solana Ruiz.

CHAPTER 9

Quantum Aesthetics and the Polity

John W. Murphy and Manuel J. Caro

ORDER AND METAPHYSICS

The Western tradition has had an obsession with order. Chance had no role to play in the grand scheme of events.[1] In fact, uncertainty was thought to culminate in cultural decline and chaos. In order to avert this condition, a special version of metaphysics was introduced to ensure both natural and social harmony. In this regard, this concern for order extends to the core of the human condition. A continuous search has been underway, in other words, for a profound base that has the ability to anchor personal and collective identities. Without this foundation, life would have no justification.

The idea that persons are simply "thrown" into the world, as Martin Heidegger describes, would be disastrous.[2] Life would have no meaning or purpose; morality would have no significance other than anomie. Despite what surrealists or postmodernists think, this condition of uncertainty is not regularly viewed as productive. Disorder is not thought to support growth and development. For this reason, many trends in art and literature in the twentieth century have not received a warm reception. Their attempts to discredit publicly reason have been interpreted by most persons as uninformed and dangerous; these attacks on reason have been understood to exemplify disrespect for the principles and traditions that make possible the survival of society. Simply put, these critics were merely irresponsible pranksters.

So, what kind of metaphysics has been introduced to support order? Heidegger once described this viewpoint as an uprooted metaphysics that gradually evolved into the Western ontological tradition.[3] As should be noted, this picture looks ominous. Indeed, the aim of this approach is to create powerful preventives and remedies for disorder. A base of morality, for example, had to be introduced that would not quickly falter and collapse. This foundation had to be intimidating and durable. After all, nothing less than civilization is at stake. Something truly serious had to be invented to prevent the onset of barbarism. This force took the form of idealized cultural mandates.

At the core of this metaphysics is a philosophical maneuver that has become problematic, but necessary to provide order with the necessary gravity. To achieve this exalted status, the rationale for order had to be removed categorically from quotidian concerns. This foundation could not be plagued by the contingencies that are part of everyday life. Such a base, in other words, would have to constitute a reality sui generis,[4] or order would be replete with assumptions and other sources of bias that would gradually prove to be problematic. Ambiguity and uncertainty would slowly creep into the basic fabric of society.

This seigniorial reality had to be sequestered from considerations that could not be readily anticipated and controlled. At this juncture is where dualism enters history. From the time of the pre-Socratics, an absolute and eternal mechanism has been sought to explain knowledge and order. *Doxa* had to be transcended. Morality and Justice, for example, were thought to emanate from this source. For this reason, the designation of ontotheological was thought to be appropriate for this ontology. Nonetheless, the idea is that an all-encompassing steering mechanism is organizing everything that can be known.

At first, this principle was referred to as *nous*, *logos*, or *ratio*. Later, during the medieval period, God took the place of these considerations. And in the modern world, natural or physical laws were presumed to be regulating nature and society.

The term that is used is unimportant. What is significant is the manner in which this source of order is conceived. First, a single factor is unequivocally given the power to integrate the diverse elements of reality. And second, this apparatus is ontologically or categorically removed from these components. Employing early Greek imagery, this control element constitutes the fundamental *arché* of existence. In contemporary terminology, this description is thoroughly foundationalist and represents a "metanarrative."[5] Presupposed by this scenario is that the realm inhabited by persons is pervaded by contingencies and only by overcoming this region can stability be attained. The human element is thus an impediment to harmony and security; the human being and the related praxis constitute a flaw in the façade of normalcy and tranquility. Therefore, the thrust of the ontotheological viewpoint is to create the illusion that the maintenance of order is possible without interference from *praxis*. The resulting base of order is autonomous and divorced from the various modes of human action, such as perception and speech. In the end, the human element is sacrificed to bring order to fruition.

When translated into a social philosophy, this dualism assumed the form of realism.[6] Realists argue that society can be maintained—what they call "moral order"—only if norms are understood to transcend the human condition. Norms can be effective, in other words, only if their source is untainted by interpretation or similar modes of bias. If the mind is allowed to mingle with reality, as Émile Durkheim suggests, the likely result is anarchy, because each person will simply adhere to his or her idiosyncratic conceptions of reality.[7] As a result, society must be conceived to be a force that confronts and constrains persons. Various strategies have been followed to provide society with this authority. In the early days of modern social science, Auguste Comte and Durkheim employed various myths to elevate society in stature. Both of these writers compared society to a human body that envelops and controls its parts. Numerous other critics, who adhere to the Parsonian position, have adopted structural metaphors and describe the social world as a balanced and thoroughly integrated system. The goal of these newest approaches is to deanimate the source order and make society appear to be invulnerable. Society is comprehensive and regulated by uniform values; society, stated differently, is a substantial and oppressive force.

The point of the social ontologies, as Talcott Parsons recommended, is to avoid the internecine relations envisioned by Thomas Hobbes.[8] In this regard, control had to be uncompromising and unrelenting. But rather than introduce Greek mythology, and the accompanying speculation, society became an inert object—a body or system that is self-equilibrating. As discussed extensively by Ludwig Wittgenstein, these models are an extension of the old metaphysics that he believes has lost justification. This rendition of metaphysics is still very powerful, although the thesis that supports this general outlook has lost legitimacy. Dualism, simply put, has been challenged seriously in recent years by a variety of philosophies.

THE POLITICAL IMPLICATIONS OF REALISM

In the early 1960s, Dennis Wrong wrote that the survival of the modern world is predicated on an "oversocialized conception of man." The image has been conveyed, he argues, that without far-reaching institutional constrains, persons will run amok.[9] Of course, this proposal is inherited from Comte and Durkheim and offers a very conservative vision of society. This outcome is particularly disturbing in a society that hopes to be democratic. After all, in this sort of polity persons are supposed to be free and self-governing. But in a Hobbesian universe, unfettered action such as this is not feasible. In the absence of widespread and unrelenting constraint, society will collapse.

As a result, the politics that are linked to realism are quite dismal. For example, Erich Fromm believes that the associated polity is engendered through self-denial, authoritarianism, and sickness.[10] To ensure tranquility, constraints must be constantly improved and expanded. A significant by-product of this strategy is self-doubt and withdrawal from the public sphere. Additionally, personal or collective plans are equated with irrationality and become understood as threatening to society. In this sense, persons are alienated from themselves and lack the tolerance, fortitude, ingenuity, and other traits necessary to act in a democratic manner. The citizenry is merely a residual category.

The Individual

According to the systems model, for example, individuals are eviscerated. In fact, proponents of these schemes do not even talk about persons. Instead, persons are synonymous with roles or some other structurally imposed parameters. The raison d'être of citizens thus comes to be identified with these structural imperatives. For these reasons, many European writers have argued that this approach represents the most cynical theory yet proposed to describe the operation of society. Persons, in short, are helpless and unable to do anything but conform to a system of norms and rules.

Indeed, according to this scenario, persons are considered to be directionless and useless until they have begun to adopt their prescribed roles.[11] Most problematic about this process is that all rational thought is imparted by the social system. Any information that does not emanate from this source is treated as unreliable and quickly marginalized. As Parsons was fond of saying, persons represent nothing more than useless passion until they are given meaning and direction by society.[12]

But what about social criticism? Democracy is supposed to be anathema to ideology. Yet, if citizens only know what the social system imparts, the element of critique is undermined. And if the system happens to counsel racism, sexism, or some other from of discrimination, these antidemocratic practices cannot be overcome. Impediments to democracy cannot be reflected on and demonstrated to be untenable, unless the system evolves and begins to reveal its contradictions. Such turn of events, however, is not very likely when the source of order and morality is autonomous and impregnable.

The Social System

What persons begin to encounter, accordingly, is what Herbert Marcuse calls an "affirmative culture."[13] Because the social system is the locus of reason and conformity, most persons are relegated to the periphery of the political process. Those who represent the core of society most closely, therefore, become the centerpiece of the polity. These select persons often come mostly from the dominant economic and cultural classes because of their close association with the cultural ideals that are thought to unite society.

What occurs in political practice, accordingly, is a battle for the soul of a society. Those individuals or groups who embody the ideals that are expected to be affirmed by everyone must disparage and banish the remainder of the citizenry to the margins of society. These exalted persons have reason, culture, and the future of humanity on their side. A natural or essentialist hierarchy is presumed to exist that should not be violated, or society may be brought to the brink of ruin. Clearly, democracy is not an important part of this perspective.

When this affirmative culture becomes ingrained, a set of so-called basic values is identified. These may relate to literature, renditions of history, or personal characteristics. In any event, these facets of culture must be differentiated from the rest and internalized by everyone or society will digress. As described by Mathew Arnold, these elements of culture are extolled as the best that humans have ever created.[14] The accompanying "cultural wars" are therefore anything but frivolous; indeed, truth, moral standards, and the security of the public hang in the balance. Any contamination of these cultural ideals is a threat to humanity.

Disembodied Culture

Because society is autonomous and centered around ahistorical ideals, the result is the pursuit of normalcy. In fact, the survival of society is thought to depend on the reduction of possible cultural differences. Every institution must have a set of core values that are given precedence over all other options, or order will be difficult to maintain. In the end, cultural homogeneity is fostered; everyone is expected to be integrated snugly under an umbrella of common themes and expectations.

Rather than oppressive, this outlook is thought to be liberating. Usually, the required adjustment has been referred to as assimilation.[15] The idea is that everyone should abandon their ethnic heritage, or other aspects of their ethnic or personal histories, and acquire a new common identity. This identity, moreover, represents an improvement over an ethnic past that stifles mobility and success in the modern world. Old family patterns, community loyalties, or customs, for example, seem to slow the reactions that are necessary for persons to act rationally and profitably at the market place. Therefore, those who are smart strive to eliminate these so-called rough edges from their appearance or character. Clearly, achieving normalcy has become a worldwide obsession. Through cosmetics, fashion, and even surgery, many persons try to revamp themselves daily and acquire a new persona. Despite the prevailing ideologies about success and mobility, Fromm describes these societies as infirm. They are ill because the subtle message is that intolerance is not only acceptable but also profitable. Because those who do not conform threaten the smooth operation of the entire social system, and thus everyone's advancement and happiness, a very narrow range of behavior must be instituted. Marcuse, for example, refers to such a society as "one dimensional"[16] and reductionistic.

Consider the recent reaction to multiculturalism. Even liberals have conceded to conservatives that this movement jeopardizes the vitality of modern societies. In this case, the principle that every culture represents particular beliefs and commitments, which should be treated equally, becomes a threat to order. As a result, acceptable cultural differences are manufactured by the advertising industry, while real persons who exhibit diversity are attacked around the world almost everyday in the streets. The message is clear: those who create themselves and exhibit authentic cultural differences are a social liability. Particularly important, they confound the pursuit of normalcy by taking pride in their uniqueness and demanding to be recognized in their own terms.

Political Action

Due to the pervasiveness of realism in the dominant social imagery, political action is severely truncated. According to realists, an objective framework—in the form of values or structural imperatives—dictate behavior. And as part of the assimilation process, persons are expected to mimic these constraints. Normalcy is attained, moreover, by internalizing these ideals. As should be noted, key to the success of this process is that the source of order is externalized. When translated into political terms, this facet of realism supports a very conservative approach to politics.

What is extolled under the guise of realism is a politics of pragmatism. Reasonable politicians, in other words, do not violate the institutionalized claims that constitute reality. They examine the present social conditions and work within the prevailing traditions—economic, cultural, or political. Given this tendency, the worst name a politician can be called is idealistic, because idealism is synonymous with irresponsibility, lacking maturity, and, in general, being out of step with the times. How could this kind of idealistic person be trusted?

Although these politicians may be predictable and trustworthy, what kind of proposals can they be expected to advance? Mostly, they will support actions that are possible. But what social problem was ever solved with this attitude! During 1968 in France, for example, people began to promote a politics of the impossible that reflected what persons could dream and accomplish. After all, why be intimidated by reality! Nonetheless, realism does not support this sort of audacity, confidence, and rebelliousness. Realists assume that conformity leads to progress. The result of realism, however, is that reality is equated with the status quo. Some critics claim that this association was intended all along; what exists currently is presumed to be the best of all possible worlds. History has come to an end. And because there is nothing remaining to say or do, the same political options are continuously replayed. There is literally a very narrow space within which persons can maneuver.

But perhaps the prevailing social system is the source of problems? What if certain structural themes have promoted the reification of barriers to advancement or equal treatment? For realists, these questions are indicative of an absence of pragmatism and reason and signal an unwillingness to compromise. However, such intransigence may impede the discovery of solutions to social problems. Real solutions may require the adoption of a new reality. And unwillingness to become unrealistic may make politics irrelevant and condemn many persons to living dismal lives.

THE QUANTUM WORLD

As should be noted, order has been predicated on a dubious assumption. The old metaphysics rests on the claim that the human presence can be overcome, and thus existential contingencies can be avoided. A space is thus available where phenomena such as Durkheim's "reality *sui generis*" and Parson's "ultimate reality" can reside.[17] This space, accordingly, provides a perspective from nowhere—an eternal vision unfettered by quotidian concerns that may compromise objectivity.

But subsequent to the advent of quantum science, this special domain is no longer available. What Heisenberg recognized was the fundamental contradiction in proposing the existence of an eternal version of space. Such a pure space, in short, could never be known or described. This kind of space had to be basically analytical and thus represent an idealized form. In a more concrete sense, this abstract space is nothing but a mathematical projection.

In order to avoid this paradox, quantum writers did not ignore the existence of human beings. These authors understood that all knowledge is mediated by the human presence, and therefore space could never be pure extension. Space had to be viewed from here or there and represent a particular way of opening the world. They recognized, therefore, that the human element had to be part of any equation that is proposed to describe nature or social life.

This association between persons and reality is integral to understanding the "indeterminacy effect." Whatever is known, in short, is shaped by human interpretation.[18] In more symbolic terms, "A" could be "B," "C," "D," and so forth, depending on the perceptual field in which "A" is placed. Every identity, as Charles Jencks notes, is "double-coded."[19] This location or field, moreover, is inextricably tied to the way in which humans are introduced. For example, their placement, speed, and direction will determine the shape of space. Space and perspective are not separate considerations any longer.

Contrary to what Durkheim recommends, in a quantum world facts are not things.[20] They cannot be deanimated in this way; they are not inert. Now, facts have a human texture, an existential core that cannot be overcome. Following the onset of this viewpoint, everything must be known from either here or there. Given the ubiquity of human presence, the type of disinterested investigation presupposed by the old metaphysics represents a severe distortion of the learning process.

Advocates of quantum aesthetics reject social realism.[21] They argue that the philosophical maneuver necessary to substantiate this outlook is no longer tenable. Cartesianism, in other words, has become difficult, if not impossible, to justify. While agreeing with the principles of quantum physics, Gregorio Morales argues that all knowledge resides in "morphogenic fields" (*campos morfogenéticos*).[22] His point is that no knowledge is pure and devoid of values; stated differently, everything that is known exists within a scene that is created by human presence. This realm, as might be expected, pulsates constantly due to the effects of personal and collective biographies.[23]

As a result, Jean François Lyotard writes that all knowledge is "locally determined."[24] Facts, for example, are understood to be embedded within the movements that are instigated by human action. Knowledge, therefore, is thoroughly sullied by the various modalities of human existence, such as passion, fear, or pain. For this reason, many contemporary French writers use both poetic and scatological language to describe the acquisition of knowledge. Truth, for example, has been described as tied to orgasms, bodily fluids, and sensuous embraces. Morales also links eyes, body, and eroticism to reality. In general, the purpose of this terminology is to demonstrate that knowledge is a human product, rather than something that is enclosed within an idealized sphere.

In this regard, matter and spirit are joined in estética cuántica.[25] Although consistent with quantum physics, this mixture violates the basic premise of Western religious and intellectual traditions. Throughout history, spirit had to be pristine and a reliable source of inspiration, honor, and truth. In the hands of quantum writers, however, spirit is worldly and thus themes that were formerly idealized now have social and cultural limitations. In the most positive sense, these notions have been given a material (or existentially based) identity. Quoting Arthur Eddington, Morales declares that the world comprises "mental" or thinking matter.[26] The mind, therefore, does not simply confront but significantly influences the perception of what comes to be known as real.

Key to estética cuántica (quantum aesthetics) is that the old metaphysics is defunct. Accordingly, anything that relied on this philosophical viewpoint must be reconsidered. Due to the collapse of dualism, there is no ultimate telos; according to estética cuántica, the historical trajectory is inaugurated from within history. And politics is not exempt from this transformation. But as can be imagined, the traditional politics of realism and control has lost its justification. Simply put, in a quantum world there does not exist an absolute foundation to subdue persons and restrict their freedom. The reality sui generis that has usually performed this function must emerge from political practice, or from the people who were once thought to be merely the source of error and discarded.

THE DECENTERED POLITY

Nowadays, many critics have begun to argue for a decentered polity. The implication is that the current approach to governance has become too entrenched, and thus a variety of persons are systematically excluded or marginalized. An important part of this critique is that social change is necessary and should be immediately undertaken. But what does decentralization actually mean? Clearly, merely redistributing power, along traditional political lines, is insufficient to guarantee that another hierarchy will not emerge. After all, a polity indebted to realism is established on a social ontology that encourages the division of society into a center and periphery.

Perhaps a more profound, philosophical maneuver is required to decentralize thoroughly the polity. At this juncture is where estética cuántica is crucial. Due to the unequivocal dismissal of dualism proposed by this theory, society cannot be legitimately recentralized. Following the collapse of dualism, no rendition of norms or values can be presumed to exist sui generis. As a result, there is no Archimedean point available that can be invoked to serve as a universal base of order, around which all persons can coalesce. In this sense, order has disappeared.

In the sense intended by Jean Baudrillard, society has vanished in several ways.[27] First, order is not something that is greater than humans; order, in other words, does not have the autonomy necessary to control persons. And second, citizens must manage themselves, rather than rely on political strictures for guidance. No longer do so-called natural political forms have the requisite power to dictate the course of events. Now, persons must reinvent themselves as a polity, without the assurances provided by political a prioris.

Most problematic about this proposal, according to most conservatives and some liberals, is that the persons who were formerly in need of control must create political reality. As a result, order is established on a source that is uncertain and unreliable. With the demise of the old metaphysics, however, there is another option; perhaps a choice does not have to be made between rationality and irrationality. If there is no refuge from praxis, the traditional absolutes are not as unambiguous and universal as was once thought. Therefore, rejecting them does not necessarily signal the absence of reason and discipline. Challenging traditional modes of thinking about the polity, for example, does not necessarily represent a loss of reason but the development of new forms of rationality. And why should persons begin again to engage in the self-denial that is essential to endow these ideals with their ahistorical status and appeal?

Maybe such legerdemain is not needed to produce a polity that is trustworthy and socially responsible. After all, a terrible prejudice has been perpetuated by the old metaphysics, one that has precluded the possibility that persons can truly manage themselves. Surely, they can invent long-term constructions, adhere to these principles, and create a process of managing change? Regardless of the answer, in a quantum world there is no other choice—persons must assume the awesome and liberating responsibility of self-control and self-governance. Now, self-legislation is possible without automatically courting chaos due to the collapse of dualism.

NOMAD SOCIETY

In the space left behind by the collapse of the old metaphysics, society can be neither centered nor recentered. Now, every direction is an option or fragment that must seek justification. And this legitimacy must emerge from the ambiguity that is difficult to quell within the space of interpretation where quantum writers and artists work. The inauguration of the polity can thus begin anywhere and spread in any direction. As Jean Gebser describes this state of affairs, the "center is now everywhere."[28] A center is now a matter of choice and includes the activities that reinforce this option, rather than an unfettered location.

The term "nomad" is popularized by Gilles Deleuze to characterize the world subsequent to the disappearance of the old metaphysics.[29] Nomads are not trapped or restricted in any way; they make their own destinies through uncharted and foreign lands. But what prevents the world of nomads from becoming chaotic? Can a polity be held together without the privileged realities that were supported by the old metaphysics?

Consistent with quantum physics, proponents of estética cuántica answer in the affirmative. They remind their readers that initiative and purpose are significant parts of the embodied world of quantum physics. Remember that the world is seen and opened from a particular vantage point; the world is given meaning and form by perception. As a result, a telos is present, but not the one that has been thought to guide history from Aristotle, through Georg Hegel, to modern scientific writers. This telos, in short, is not abstract but purposeful; this purpose is not ethereal but worldly. On the basis of this internal telos, so to speak, a socially responsible polity can be established.

Although this worldly telos has a variety of manifestations, only a few can be dealt with in this chapter. Most important about them all, however, is that they flow from praxis, are subject to reinterpretation and redeployment, and occupy a space that is not indifferent to its contents. As a result, this world can have a purpose and be integrated without the universals that are usually invoked to ensure order. Any phenomenon that exists in this way, accordingly, is thoroughly worldly and subject to human creativity.

The New Individual

In this new, nomad polity, essentialism is passé. Persons can no longer be viewed to have essential or fundamental traits that determine their respective identities, worth, social positions, and so forth. As should be noted from the previous discussion, this variation of foundationalism cannot be justified. After the collapse of the old metaphysics, there does not exist a fundamental core that anchors an individual to a particular biographical moment. Now, biographies are much more intimate, unique, alive, and variegated.

Furthermore, essentialism is anathema to a democratic or open polity. Throughout history, essentialism has established strict parameters with regard to who possesses a solid character, good ideas, or the right to express demands. If a polity is predicated on essentialism, hierarchy and marginalization can be expected. From the time of Plato, essentialism performed this function.

In a nomad polity, persons should not be labeled in an exclusionary manner, either positively or negatively. The reason for this conclusion is simple: Persons have no other choice but to make themselves in one way or another, without the luxury of having their personal projects legitimized by founding principles or guidelines. As Morales likes to say, becoming an individual is an obligation. Persons, as he says, can rebel against their destinies.[30] Identities, therefore, are concocted or made up throughout life and are influenced by the intersection of many factors. And eternal standards of inferiority or superiority, for example, are impossible to discern from this myriad of considerations. Portrayed in this manner, these designations cannot retain the patina of purity or universal-

Quantum Aesthetics and the Polity

ity that they require to retaining their power. Without the influence of dualism, nothing about an identity is clear-cut or obtrusive.

According to Morales, identities are potentially infinite.[31] The limits that emerge through the process of identity formation are a product of persons managing their existences. In a quantum world, identities are not merely the result of elements that reside either inside or outside of the person, such as the subconscious (not necessarily unconscious) or social rules. Now, praxis pervades identity; every factor that might affect personal growth must be incorporated and (re)evaluated before it has any relevance. In the Jungian language found throughout estética cuántica, persons "animate" themselves.[32] Additionally, this praxis is both individual and collective. Most important is that identities are not inherent and can be regularly redirected. Moral character, moreover, is not an indigenous property of select persons, but is a trait that develops in various ways through many means.

Considering the fluidity of identity, Morales refers to the process of individual growth as "individuation." He borrows this idea from Carl G. Jung, but does not include the dualistic and static metaphysics that are often attributed to the famous psychologist. For Morales, a person is never uniform but rather diverse and complex.[33] Referring to a person as having an identity is incorrect; every identity, instead, is multiple and mixed. Furthermore, notes Morales, this is precisely the sort of diversity that is undermined by social realism and the accompanying abstract versions of the polity.

The usual justification for systematically excluding certain persons from the polity is now defunct—an unequivocal mechanism for distinguishing and inferiorizing persons is not sustainable. What a person has to say must be tested by debate and close examination, instead of linked to natural propensities and evaluated on the basis of these principles. For example, the politics of "pigmentocracy" made famous in the studies by Frantz Fanon has been rendered inoperable by this critique of essentialism.[34] Following the onset of estética cuántica, a nomad polity is possible. That is, participation in the process of governing can expand and viewpoints can proliferate because no persons or positions are tainted by insuperable flaws or limitations. Hence, each segment of the population deserves to be consulted and taken seriously; none is, after all, fundamentally more valuable than any other. In the public space that is now available, any differences that exist can be assessed, and possibly winnowed, through discussion rather than through the exercise of bias and prejudice that is rationalized by invoking nature or science. These kinds of natural or automatic justifications have lost their authority.

The New Social System

Many writers believe that essentialism is not all bad. Having a foundation that anchors society provides a vital function—that is, order is given a sound rationale. Moreover, persons are provided with a clear picture of where they fit into the social system. In this regard, a hierarchy imposes discipline and the control that is thought to be necessary for society to survive. The central problem with this methodology is that the sacrifice of diversity and freedom is expected and encouraged.

But does the survival of order require this sort of trade-off? Advocates of estética cuántica do not believe so. In fact, they are constantly championing the traits of individuation and diversity. According to the members of El Grupo, a world that is tolerant of diversity is one where humanity can prosper. Morales even goes so far as to say that estética cuántica and diversity are synonymous.[35] And with such a progressive focus, the exercise of liberty is enhanced.

But what about disorder? With such a strong emphasis placed on individuation, surely the risk increases that disorder might erupt. The Hobbesian problem, as Parsons called it, might become reality. Nonetheless, remember that these writers are not dualists, and therefore individuation is not necessarily the same as individualism.[36] The individual is not an atom that must be joined with similarly isolated elements through the force exerted by an overarching social superstructure.

Consistent with antidualism, order is not arranged in a vertical manner in estética cuántica. Order is arranged directly, instead, through the "integration of opposites."[37] In this case, the issue is not simply that "A" could be "B," but that society and the individual are part of the same process.[38] Accordingly, Morales notes clearly that becoming individuated involves criticizing and borrowing many elements from the community. As a result, the biographies of the individual and community are intimately intertwined. Clearly, the one presupposes the other; the individual and community are not antipodes.

In terms of promoting a nomad polity, this link is very important. First, a polity can be engendered without the influence of social realism and the resulting hierarchy. Second, and especially relevant to promoting open discussions, every personal choice is tied to other modes of individuality. Morales makes this point by saying that the part contains the whole.[39] No cultural difference, therefore, resides within a world of its own; identity and difference presuppose each other and exist together. Politically, this understanding is vital to producing the tolerance and tranquility that was thought to be forthcoming only from social realism.

No political position exists in isolation from others. But the social world, in a nomad polity, is not a free-for-all. Instead, political positions are expected to consult with one another regularly, due to the fact that they share a common world and destiny. They cannot escape from or deny their coexistence. And as mentioned earlier, this consultation must never involve domination. In a nomad polity, every act is a coaction that should not violate the presence of others. As Martin Buber described, the "I" and the "Thou" constitute the type of copresence that entails mutual respect and toleration.[40] Only when persons look to heaven for salvation, writes Buber, do they lose sight of their compatriots. Only when persons are understood to rely on a reality sui generis to engender order can they be treated as if they are disconnected from each other's acts.

Aristotle once said that humans are fundamentally political animals. Supporters of estética cuántica agree with his claim with one major caveat. The polity, in short, has nothing to do with the state. The state is simply too rarefied and organizes persons from a distance. With regard to estética cuántica, however, the polity arises from action that is unrestrained except by the presence of others. This "difficult freedom," as labeled by Emmanuel Lévinas, is all that is possible in a nondualist world where the self and other share an existential domain.[41] Freedom is thus always implicated in the presence of others.

For some persons, however, this tie represents an unwarranted barrier that is placed on their freedom. The problem is that they believe they have the infinite freedom associated with an atom. This vision, nonetheless, represents an old world where dualism made sense and was accepted. In a quantum world, infinity is always concrete or bound to a particular time and place, and freedom has the same constitution.

Most unfortunate is that a powerful economic ideology such as capitalism clings to an outmoded image of existence, and continues to encourage persons to reach eternity through their personal acts of consumption. As a result, in many places around the world morality is touted to be personal, and thus the interpersonal sphere of existence—the realm "in-between" the I and other—collapses.[42] Hence, society becomes a collision of atoms that have the basic right to do whatever they please. In this dualistic world, freedom is little more than perverse license.

Political Action as Praxis

In the quantum world, to use Jean-Paul Sartre's phrase, the polity represents "collective *praxis*."[43] Within this framework, persons are understood to open the world in unique ways. In some Latin American countries, this view of the polity has come to be called "mature democracy."[44] The point is that the necessary philosophical maneuvers have been made to discourage domination and increase civic participation.

In the absence of realism, persons can develop the world they envision. Since the polity exists at the level of interpersonal constitution, order does not confine persons to any particular fate. In this sense, they never confront their destiny, but instead make proposals that are neither realistic nor unrealistic. As Friedrich Nietzsche discussed with regard to good and evil, this dichotomization of reason has lost significance.[45] Now, the value of a proposal is discovered in praxis, or in the desire of persons to make a particular option real. The politics of pragmatism thus gives way to the ability of equal coparticipants to make a new reality that earlier may have been considered utopian.

Despite what some critics may think, this image of a quantum polity is not utopian. No claims are made about the complete absence of conflict or the prospects for unfailing cooperation. What this position reveals is that there is no ontological justification for asymmetry between persons or groups, unlimited freedom cannot be rationalized, and history is unfinished and can assume any humanly inspired direction. While not utopian, this polity allows persons to grow together in many ways. A commitment to a particular course of action, therefore, gives a particular articulation of reality a sense of purpose and urgency.

Morales argues in a similar manner when he states that humans occupy the center of the universe.[46] This claim is not merely another attempt to resurrect the usual rendition of anthropocentrism; his aim is not to restore humans to the top of the evolution scale. Instead, and much more profound, Morales is acknowledging the basic tenet of quantum physics: humans give shape and purpose to whatever they know. Their influence is undeniable.[47] For the members of El Grupo, the political implications of this epistemology are obvious. That is, political realities are made and not simply encountered.

Political pragmatism, according to this viewpoint, has nothing to do with reason or prudence. Those who are pragmatic are simply committed to a particular reality or lack the will or courage to pursue another direction. In the quantum world, neutrality is not possible and is exposed to be a façade; persons do not simply watch history unfold. Moreover, the question must be asked, how can the privileged status of an unencumbered reality be sustained? Quite simply, this rarefied position is affirmed through a powerful ideology that is undermined by estética cuántica. Political necessity, in other words, is rationalized by a reality that humans do not create but must internalize. Clearly, quantum writers have exposed this chicanery and self-denial that is at the root of this sort of pragmatism. And of course, in some circles, this denial is considered to be indicative of alienation and a sick society.

CONCLUSION

Although not developed intentionally as a social philosophy, estética cuántica has important political implications. Indeed, the inability to sustain realism requires that alternative political strategies be sought. The members of this movement have recognized this need and have begun to describe the modes of freedom and politics that are possible. With the old metaphysics under siege, the resulting democracy must grow from below and represent the diversity that is unleashed. These different manifestations of praxis, moreover, are integrated in such a way that their unique qualities are maintained.

As should be noted, this outlook is not necessarily consistent with recent trends to democratize societies. These new opportunities have tied democracy to a market mechanism and the accompanying liberal state. Clearly, the importance of estética cuántica extends much farther than such a system in two significant ways. First, individualism is not equated with egoism and the right of persons to pursue their own aims, without regard for the presence of others. From the perspective of estética cuántica, this market scenario breeds unhealthy rivalries and supports privileges that eventually contravene democracy. Those who support the market as the general palliative of social troubles do not usually raise issues related to political power and other sources of discrimination.

And second, liberal states tend to centralize power in such a way that order is easily reified and the exclusion of groups becomes common place. In estética cuántica, no factors remain that have the status to justify these concentrations. Because the state and similar abstractions are not needed to organize persons, unassailable systems and dirempt institutional arrangements do not have to be preserved simply to calm fears about disorder. Abandoning these arrangements and strictures does not signal the collapse of society, but merely the opening of alternative worlds. What is clearly illustrated by estética cuántica is that persons do not have to sacrifice themselves to preserve order.

In this sense, estética cuántica takes another step toward the thorough democratization of culture that is necessary to establish a democratic polity. The focus is not simply on opening additional mechanisms of government, but on establishing the philosophical principles that are required to promote and institute full civic participation in all facets of society. Therefore, as defined by Karl Marx, estética cuántica goes to the root of democracy. The basic framework of democracy is addressed, instead of merely some procedural shortcomings. This shift in orientation has always been considered quite radical, because various institutionalized assumptions about economics or race, which have served to justify inequality, are now available for critique. Given this level of analysis, current democracies may be significantly improved by estética cuántica.

NOTES

1. John W. Murphy, *Postmodern Social Analysis and Criticism* (Westport, CT: Greenwood, 1989), 1–19.
2. Martin Heidegger, *Being and Time* (New York: Harper and Row, 1962), 223 (see also 179).
3. Ibid., 42–49.
4. Émile Durkheim, *Pragmatism and Sociology* (Cambridge: Cambridge University Press, 1983), 85.
5. Jean-François Lyotard, *The Postmodern Condition* (Minneapolis: University of Minnesota Press, 1984).
6. Warner Stark, *The Fundamental Forms of Social Thought* (New York: Fordham University Press, 1963), 1–13.
7. Durkheim, *Pragmatism and Sociology*, 23.
8. Talcott Parsons, *The Social System* (Homewood, IL: Free Press, 1951), 36.
9. Dennis Wrong, "The Oversocialized Conception of Man in Modern Sociology," *American Sociological Review* 26 (1962): 183–193.
10. Erich Fromm, *Man for Himself* (New York: Rinehart, 1958), 68–82.
11. Talcott Parsons, *Societies* (Englewood Cliffs: NJ: Prentice-Hall, 1966), 29.
12. Ibid.
13. Herbert Marcuse, *Negations* (Boston: Beacon, 1968), 95.
14. Mathew Arnold, "The Function of Criticism at the Present Time," in *The Portable Mathew Arnold* (New York: Viking, 1965), 234–267.
15. John W. Murphy, and Jung Min Choi, *Postmodernism, Unraveling Racism, and Democratic Institutions* (Wesport, CT: Praeger, 1997), 16–19.
16. Herbert Marcuse, *One-Dimensional Man* (Boston: Beacon, 1964).

17. Murphy, *Postmodern Social Analysis and Criticism*, 58–60.
18. Werner Heisenberg, *Physics and Philosophy* (New York: Harper and Row, 1962), 44–58.
19. Charles Jencks, *The Architecture of the Jumping Universe* (London: Academy Editions, 1993), 168–169.
20. Émile Durkheim, *The Rules of Sociological Method* (New York: The Free Press, 1982), 60–89.
21. Gregorio Morales, *El Cadáver de Balzac* (Alicante: Epígono, 1998), 14.
22. Ibid., 38.
23. Ibid., 65.
24. Lyotard, *The Postmodern Condition*, 61.
25. Morales, *El Cadáver de Balzac*, 19.
26. Ibid.
27. Jean Baudrillard, *In the Shadow of the Silent Majorities* (New York: Seniotext(e), 1983), 66.
28. Jean Gebser, *The Ever-Present Origin* (Athens: Ohio University Press, 1984), 544.
29. Gilles Deleuze, "Nomad Thought," in *The New Nietzsche*, ed. David B. Allison (New York: Dell, 1977), 141–149.
30. Morales, *El Cadáver de Balzac*, 27.
31. Ibid., 25.
32. Ibid., 47.
33. Ibid.
34. Franz Fanon, *The Wretched of the Earth* (New York: Grove Widenfield, 1991), 41 ff.
35. Morales, *El Cadáver de Balzac*, 35.
36. Quantum Aesthetics Group, *Quantum Aesthetics Group's E-m@ilfesto*, 9 February 2000, ⟨http://teleline.terra.es/personal/lucschok/estetica/emailfestoeng.htm⟩.
37. Ibid.
38. Morales, *El Cadáver de Balzac*, 25.
39. Ibid., 41.
40. Martin Buber, *I and Thou* (New York: Scribner's, 1958).
41. Emmanuel Lévinas, *Difficult Freedom* (Baltimore, MD: Johns Hopkins University Press, 1990).
42. Buber, *I and Though*, 11.
43. Jean-Paul Sartre, *Critique of Dialectical Reason* (Atlantic Highlands, NJ: Humanities, 1979), 505–524.
44. For example, see Alejandro Serrano Caldera, *La Unidad en la Diversidad* (Managua: Editorial San Rafael, 1993), 123.
45. Friedrich Nietzsche, *Beyond God and Evil* (Amherst, NY: Prometheus, 1989).
46. Morales, *El Cadáver de Balzac*, 38.
47. Ibid., 77.

CHAPTER 10

Conclusion: The Renewal of Cultural Studies

John W. Murphy and Manuel J. Caro

INTRODUCTION

Quantum aesthetics is part of a movement toward an intimate metaphysics initiated after 1900. Existentialism, phenomenology, and various artistic traditions belong to this trend. Contrary to the traditional outlook, proponents of quantum aesthetics have illustrated that everyday life is metaphysical and worthy of appreciation, just as Max Horkheimer and other writers have done.[1] As should be noted throughout this collection, however, this metaphysics is not weighty, and thus is neither the source of, nor inspired by, fear and trembling. Instead, this metaphysics is embraced by persons and sustained by their daily concerns.

In this sense, metaphysics is no longer foreign. More recent theories, such as postmodernism and its variants, have tied metaphysics to language and transformed reality into something linguistic and insubstantial. To paraphrase Roland Barthes, even objectivity—the most highly praised and valuable source of knowledge—is simply another rendition of human praxis.[2] In quantum language, there are no "hidden variables."[3] Because this mode of inspiration cannot be overcome, so-called subjectivity extends indefinitely. And with this mode of interpretation linked to metaphysics in this way, the products of human action are never finished and always being rewritten.

This openness that has been prompted by the humanities has also been a source of criticism. Stated simply, a general sense of metaphysical intimacy has not had widespread acceptance. Reluctantly, some persons have begun to recognize the elusive character of truth, but only in the humanities. The reality conveyed by a short story, for example, may be mediated by conceptual schemas related to class or gender. Nonetheless, as readers are constantly reminded by realists, the physical side of existence is not plagued by this ambiguity. Nature, in other words, has laws that are imposed on everyone without exception.

At this point is where quantum aesthetics enters the scene. This philosophy of art reveals that interpretation is not restricted simply to culture. According to supporters of quantum aesthetics, interpretation is not a cultural anomaly but pervades the deepest recesses of nature. Borrowing from Werner Heisenberg, these artists contend that perception and cognition play a key role in shaping and reinforcing the worldview that comes to be known as real.[4] Nothing escapes unscathed by the human element, even factors that appear to be most objective.

Quantum aesthetics fosters what might be called a general thesis of metaphysical intimacy.[5] There is no place left, even in nature, where uninterpreted events can hide. With regard to the work of Niels Bohr and Heisenberg, this condition of unavoidable interpretation is referred to as the "indivisibility of the quantum action."[6] Accordingly, talking about any privileged or pristine considerations involves contradictions that, according to advocates of quantum aesthetics, must be overcome. Now, every facet of existence has a voice that has a human origin and must be correctly deciphered. Even nature is not simply confronted, but must be explored in terms of its cultural nuances.

The question that remains is: How does this shift in thinking affect the field of cultural studies? Most important is that the old dualisms cannot be sustained anymore, for they require that important cultural factors be sequestered from interpretation. This chicanery is undermined by quantum aesthetics. For example, the usual distinction that is made between method and fact loses its viability. Therefore, rather than guided by the search for objectivity, the embedded character of facts is the focus of cultural studies.

This conclusion does not mean that this style of study lacks rigor or depth. Rather, the point at this juncture is that the pursuit of uninterpreted facts may sound impressive but makes no sense. Following the onset of quantum aesthetics, cultural studies is not merely an approach adopted by those who are deluded or lack the skill or talent to become real scientists. Because everything is embedded within interpretation, science is also cultural. Those who give credence to cultural studies recognize the ubiquity of culture and openly acknowledge that becoming acultural is impossible. All research is thus existential or cultural. This new research is also based on the overcoming of a series of dualisms, some of which are discussed in this chapter.

Knowledge and Praxis

The usual distinction that is made between praxis (subjectivity) and reality (objectivity) has been invalidated by quantum aesthetics. As should be noted, however, praxis is not the same as "selective perception." In other words, praxis involves a lot more than choosing among already constituted options;

Conclusion 183

objectivity is not simply "disturbed" by the human presence, but is known only in terms of this productive force. Through the exercise of this human, creative capacity, persons are able to make and continuously remake reality. Whatever is recognized as real is no longer autonomous, but rather is supported by imagination and desire. Reality is nothing more than a specific cultural construction that receives special attention.

Method and Fact

Usually, method has been viewed as a neutral device that provides researchers with access to facts.[7] But now, method is also an outgrowth of praxis; method is a particular mode of praxis, and is not merely a conduit for information. In a manner of speaking, through the use of methodology the type of data to be collected is specified. So, what can be done so that the voice of data is not muffled by methodology?

Most important is that methodology should be approached as a means of communication, rather than a strategy for data collection. Collection is disinterested and sterile, while communication requires intimacy and cultural relevance. Therefore, methodology should be designed to let data speak in their own voice, thereby avoiding reductionism. As a mode of inquiry, methodology should be made to reflect the style of praxis of those who are studied. In this way, the integrity of data is maintained and social reality is not obscured by irrelevant methodological maxims.

Intimacy, in view of quantum aesthetics, is not something that contaminates data. Given Heisenberg's indeterminacy principle, intimacy is present whether or not researchers want to acknowledge its effects.[8] All advocates of quantum aesthetics want to do is ensure that the benefits of this intimacy are recognized and utilized. Cognitively imposed presuppositions are present invariably throughout research. However, quantum investigators want to bring to the forefront those factors that let people express their world in their own way, while disposing of those that may obscure people's praxis.

Individual and Society

Traditionally, a clear distinction has been made between the individual and society. In fact, these elements have been treated as separate ontological entities. Many problems have arisen because of this differentiation. Most noteworthy is that having social responsibilities is thought to impede the growth of individualism. Furthermore, on the policy side, the collective good is thought to be fostered only by encouraging the realization of personal desires. The common weal is thus, at best, promoted indirectly, so that the individual is not burdened with "unnecessary" responsibilities. Persons are faced with the choice of either developing personally or exhibiting concern for the well being of others. As is very clear in today's world, this dualism has dire consequences. Often, persons reap vast personal gains but reluctantly give to charity or acknowledge any other forum of social responsibility. In quantum aesthetics, however, this fragmented image of the social world is overcome. The individual-social dichotomy is understood to be fatuous and the product of dubious political and economic motives.[9] Accordingly, preserving the social good does not automatically pose a threat to personal freedom. What quantum writers argue is that persons are not monads who are basically severed from others; personal identities are not formed, they argue, outside the interpersonal activity.

What must be accomplished, therefore, is the generation of new social imagery that does not encourage the schism between the self and the other. Some writers have argued that images such as the collage, montage, and constellation capture this new style of integration.[10] The hologram is an alternative suggested by supporters of quantum aesthetics. In each of these examples, the part and the whole belong to a single dialectical movement, with neither one having a dominant position. Clearly, this shift in thinking about how individuals relate to communities have important political consequences. For example, personal gains and social costs can no longer be viewed as unrelated. In a system such as capitalism, this conclusion will likely be resisted. Moreover, reframing social problems to encompass more than individual propensities can require far-reaching economic and other cultural changes.

Diversity and Order

Often, order is thought to constitute a reality sui generis.[11] In the Western intellectual tradition, this approach to conceptualizing order is presumed to prevent society from devolving into chaos. If the source of order is autonomous, and thus authoritative, a reliable base of order is available.

There is no doubt that this strategy can secure a firm foundation for society. Nonetheless, viewing order in this manner can be counterproductive, particularly in a democracy. Most problematic is that this style of realism encourages social homogeneity. Simply put, the norms that are touted to exist sui generis can easily become identified with cultural ideals, which every so-called normal person is expected to internalize. Failure to adopt these norms, moreover, often results in censure and marginalization.

Yet, democracy requires that diversity be fostered and protected. And with the rise of multiculturalism, this need to incorporate diversity is even more urgent. As long as norms are understood to constitute a reality sui generis, diversity will be viewed as a threat to society. Increased diversity will represent nothing more than the unwarranted proliferation of norms and the absence of a clear focus. For example, Émile Durkheim believed that this expansion would result in anomie and the eventual collapse of society.[12]

Exponents of quantum aesthetics do not harbor these fears because they are not realists. Consequently, they have never given any credence to the possible existence of a reality sui generis, and have not presumed that diversity should be quelled by cultural universals. For them, no norm can ever achieve this sort of exalted stature.

In quantum aesthetics, norms are not arranged in a hierarchy; rather, they are juxtaposed to one another and organized laterally. Norms exist side by side without the final synthesis that condenses them into a uniform system.[13] A hologram, for example, lacks this sort of synthesis. Without an autonomous reality, the presence of multiple norms can only assume the form of a constellation where order and diversity exist together. And as should be noted, adding more norms to this mixture does not threaten order. In a collage, similar to a hologram, the addition of more elements enhances the complexity and thus the beauty of the work. In these new images of society, superposition is replaced by juxtaposition and contrast as the foundations of order.

With regard to cultural studies, multiculturalism has become a very important theme. What is missing from the traditional discussions of this issue is an ethic that enables diversity and order to exist together. As described by Alejandro Serrano Caldera, this new ethic must allow unity to develop from diversity.[14] In this way, both the individual and society are enriched. Through quantum aesthetics, this merger is not only possible but also necessary. Thus, in many ways, quantum aesthetics is in the vanguard of promoting democracy by rendering new images of order that do not restrict diversity. Advocates of multiculturalism have in quantum aesthetics an argument to fight against conservative critics who declare that society cannot survive without common values and norms, or, stated differently, social homogeneity. Due to quantum aesthetics, a solution is close at hand to the old political problem of how to avoid reductionism when trying to create order. How can a society be orderly and treat fairly input from its various cultural sectors? Because of the way in which quantum aesthetics deals with the integration of order and diversity, cultural studies and the politics of diversity have been significantly advanced.

Regulation and Determinism

Most social sciences are enamored of becoming truly scientific. The goal of these various disciplines, accordingly, is to discover the causes of behavior. If these key variables are unveiled, the belief is that laws of explanation can also be formulated. Behavior can thus be predicted and controlled with relative ease. What more can be expected from a natural science?

According to this scenario, causes are autonomous and prompt persons into action; they operate as stimuli that have inherent properties. Because these elements are universal—unaffected by context and other situational contingencies—they supply rationale for behavioral uniformity. The assumption, usually referred to as the "constancy hypothesis," is that all normal persons will respond to these stimuli in a similar manner.[15] Unaffected by interpretation, these causes supply the cement that holds society together.

Within the domain of quantum aesthetics, however, causes do not have this prominence and do not function in this way. In short, autonomous causes do not impose themselves on passive entities. In a quantum world, active factors are not tethered to passive ones to form a neat and tidy network. Because there is no absolute referent that can be used to make such distinctions, all entities move relative to one another. The impact of a so-called cause, therefore, depends on the "field" where the effects are supposed to take place.[16] In other words, natural causes are embedded within a host of contextual qualifications that mediate the significance of these stimuli.

Maurice Merleau-Ponty offers an interesting portrayal of how this finding applies to social life. Stimuli, he argues, are not imprinted on humans, but rather are deflected by human action.[17] Their importance is determined by the trajectory inaugurated by persons; their identity is dependent on the human project. As a result, causes are insignificant and divorced from the interpretive element that gives these components life. Stated differently, causes are acausal; that is, they must be constructed as causes through individual and collective expressiveness before they are viewed as relevant and able to prompt anyone into action. Factors are thus not united by causes but, as proponents of quantum aesthetics like to say, are the products of synchronicities.

A similar cause can thus generate a variety of outcomes, depending on the cultural significance of this input. Does the prospect of such variation undermine the ability of quantum aesthetics to address the issues of behavioral regularity? No. Clearly, the fact that light can be understood to be either a wave or a particle does not mean that a fairly stable social reality cannot be revealed. Indeed, the presence of consciousness does not necessarily disrupt regularities. Rather than undermine continuity, quantum aesthetics simply withdraws any guarantees that a specific worldview will remain valid forever.

Cultural studies that are inspired by this quantum philosophy are not futile. Nonetheless, their purpose is not to discover universal, deterministic laws. Quantum research is designed to reveal the conditions that specify when certain phenomena are understood to be factual and truthful. Social phenomena are neither real nor an illusion a priori. Moreover, if possible, the attempt is made to expose why these conditions have been instituted and viewed to be real. Instead of legitimizing causes, the aim is to gain insight into their social construction and justification. Given this reflexivity, however, traditional renditions of causality lose their power.

CONCLUSION: A NEW CULTURAL SCIENCE

When influenced by quantum aesthetics, cultural studies moves in a new and unmistakable direction. That is, there is no alternative for this science other than to become self-reflexive. Due to the collapse of dualism, neutrality and objectivity are impossible pursuits. After all, all knowledge is approached from somewhere or, in other words, from one perspective or another. There is no escape from this fate. In view of the theoretical heritage of quantum aesthetics, cultural studies must be thoroughly embodied, as opposed to the pristine presence coveted by positivism.

In traditional science, reflexivity is marginalized, for self-interrogation resurrects the role of values in science that can become very disruptive. Data, in short, are only considered good if they are unencumbered by the human element. However, according to quantum aesthetics, reflexivity allows for the various assumptions that shape knowledge to be explored. Obviously, the subtle or epistemological politics of science—how ordinary experiences come to be known as factual—are brought to center stage. Additionally, how science may intersect with gender or social class to become a repressive force is no longer difficult to envision. The science of cultural studies, therefore, has no option but to be socially accountable. Inspired by quantum aesthetics, cultural studies constitutes "engaged" research—investigations that unearth the political character of data.[18]

Becoming engaged has both advantages and disadvantages. Cultural studies can no longer have any pretence of advancing beyond the formulation of limited findings and explanations. Nonetheless, this traditional shortcoming may be a strength. Now, relevant data that are shaped by diverse histories and experiences may become the focus of attention. Furthermore, explanations may now be reasonable rather than simply rational; that is, they can be socially relevant instead of objective and culturally meaningless. In the end, improved data may be the result of this break with the universals that sustain normal science. A "somewhere" may be investigated thoroughly, rather than the virtual realm that is regularly the focus of typical research. This world that we are trying to avoid is virtual because it is reduced to empirical indicators that are stripped of their human significance.

NOTES

1. Martin Jay, *The Dialectical Imagination* (Boston: Little, Brown, 1973), 50–54.
2. Roland Barthes, *The Grain of the Voice* (New York: Hill and Wang, 1985), 52.
3. David Bohm and B. J. Hiley, *The Undivided Universe: An Ontological Interpretation of Quantum Theory* (London: Routledge, 1993), 2.
4. Gregorio Morales, *El Cadáver de Balzac* (Alicante: Epígono, 1998, 133.
5. Maurizio Ferraris, *La Hermenéutica* (México, D.F.: Taurus, 1998), 28–37.
6. Bohm and Hiley, *Unidivided Universe*, 13–14.
7. Ambrosio Velasco Gómez, "La Relevancia del Pensamiento de Gadamer en la Filosofía: Más allá de la modernidad y la Postmodernidad," *Theoría* 7 (December 1998), 55–56.
8. Morales, *El Cadáver de Balzac*, 107.
9. Ibid., 25–26.
10. John W. Murphy, and Jung Min Choi, *Postmodernism, Unraveling Racism, and Democratic Institutions* (Westport, CT: Praeger, 1997), 25–26.
11. Émile Durkheim, *Pragmatism and Sociology* (Cambridge: Cambridge University Press, 1983), 85.
12. Ibid., 65–68.
13. Morales, *El Cadáver de Balzac*, 189.
14. Alejandro Serrano Caldera, *La Unidad en la Diversidad* (Managua: Editorial San Rafael, 1993).

15. Frederick I. Kersten, "The Constancy Hypothesis in the Social Sciences," in *Life-World and Consciousness*, ed. Lester E. Embree (Evanston: Northeastern University Press, 1972), 521–563.

16. Morales, *El Cadáver de Balzac*, 77.

17. Maurice Merleau-Ponty, *The Phenomenology of Perception* (New York: Humanities Press, 1962).

18. Jean-Paul Sartre, *What Is Literature?* (New York: Washington Square, 1949).

Suggested Readings

Albright, Daniel. *Quantum Poetics: Yeats, Pound, Eliot, and the Science of Modernism.* Cambridge: Cambridge University Press, 1997.
Arias, Jesús. "El Artista Exhibe en Granada la Muestra 'Estética Cuántica.'" *El País (Andalucian Edition)*, 25 March 2000, 10.
Baudrillard, Jean. *Simulacra and Simulation.* Ann Arbor: University of Michigan Press, 1994.
Bohm, David. *Wholeness and the Implicate Order.* London: Routledge and Kegan Paul, 1980.
Dvorac, Mihaela. "Estetica Cuantica—o Propunere Pentru Secoul XXI." *Jurnalul Literar* (September–October 1999: 17–20.
———. "Gregorio Morales si Viziunea „Cuantica" a Esteticii." *Jurnalul Literar* (November–December 1999): 21–24.
Caro, Manuel J., and John W. Murphy. "Gregorio Morales: Etica y Estética Cuánticas." *Arizona Journal of Hispanic Cultural Studies* 4 (2000): 235–248.
Cole, Bruce, and Adelheid Gealt. *Art and the Western World: From Ancient Greece to Postmodernism.* New York: Summit, 1989.
Davies, Paul. *Other Worlds: Space, Superspace, and the Quantum Universe.* New York: Viking Penguin, 1997.
Deleuze, Gilles, and Félix Guattari. *On the Line.* New York: Semiotext(e), 1983.
Diéguez, Miguel Ángel. *En la Gran Manzana.* Alicante: Epígono, 1997.
Ffrench, Patrick. *The Time of Theory: A History of Tel Quel (1960–1993).* Oxford: Claredon, 1995.
Fortuny, Francisco. "Sobre Estética Cuántica I." *Papel Literario (Diario Málaga-Costa del Sol)* 18 February 1996, 142.
———. "Sobre Estética Cuántica II." *Papel Literario (Diario Málaga-Costa del Sol)*, 25 February 1996, 143.

García Viño, Manuel. "La novela Relativista y Cuántica: Materiales para la Construcción de una Teoría Aplicable a otras Artes." *Heterodoxia* 22 (1995).
———. *El Puente de los Siglos*. Madrid: Ibérico Europea de Ediciones, 1986.
Gebser, Jean. *The Ever-Present Origin*. Athens: Ohio University Press, 1984.
Greene, Brian. *The Elegant Universe: Superstrings, Hidden Dimensions, and the Quest for the Ultimate Theory*. New York: Norton, 1999.
Greenwood, Phaedra. "Inner Space: Artists Take a Thoughtful Look into the Universe's 'Quantum Soup.'" *Tempo—Section C (The Arts and Entertainment Magazine of the Taos News)* 20 January 2000, 8–9.
Guillén, Rafael. *Límites*. Barcelona: Colección "El Bardo," 1971.
———. "La Poesía ante el Nuevo Siglo." *República de las Letras* 60 (January 1999): 77–81.
Hawking, Stephen. *A Brief History of Time: From the Big Bang to Black Holes*. New York: Bantam, 1988.
Heelan, Patrick A. *Quantum Mechanics and Objectivity*. The Hague: Nijhoff, 1965.
Heisenberg, Werner. *Physics and Philosophy*. New York: Harper and Row, 1962.
Hoffman, Katherine. *Explorations: The Visual Arts since 1945*. New York: Icon Editions, 1991.
Ihde, Don. *Technology and the Lifeworld*. Bloomington: Indiana University Press, 1990.
Irigaray, Luce. *An Ethics of Sexual Difference*. Ithaca, NY: Cornell University Press, 1993.
Jung, Carl G., and Wolfgang Pauli. *The Interpretation of Nature and the Psyche—Synchronicity: An Acausal Connecting Principle*. New York: Pantheon 1955.
Kafatos, Menas and Robert Nadeau. *The Conscious Universe: Part and Whole in Modern Physical Theory*. New York: Springer-Verlag, 1990.
Kosko, Albert. *Fuzzy Thinking: The New Science of Fuzzy Logic*. New York: Hyperion, 1993.
Kristeva, Julia. *Revolution in Poetic Language*. New York: Columbia University Press, 1984.
Levinas, Emmanuel. *Totality and Infinity*. Pittsburgh: Duquesne University Press, 1984.
Lingis, Alfonso. *Libido*. Bloomington: Indiana University Press, 1985.
Longino, Charles F., and John W. Murphy. *The Old-Age Challenge to the Biomedical Model*. Amityville, NY: Baywood, 1995.
Lowe, Lynda. *Form and Measure (Catalog)*. Chicago: Niu Art Museum, 1999.
Lyotard, Jean-François. *The Postmodern Condition*. Minneapolis: University of Minnesota Press, 1984.
Martínez Manzano, Juan Carlos. "Xaverio: El Posible Discurso Espacialista en la Estética Cuántica." *Kylix* 5 (June–July 2000): 79–82.
Merleau-Ponty, Maurice. *The Phenomenology of Perception*. New York: Humanities Press, 1962.
Morales, Greorio. *El Cadáver de Balzac*. Alicante: Epígono, 1998.
———. "Ciencias Más Letras." *República de las Letras* 60 (January 1999): 97–100.
———. "Cirlot Cuántico." *Por Ejemplo* 9 (April–September 1998): 27–41.
———. *La Cuarta Locura*. Barcelona: Grijalbo, 1989.
———. *El Devorador de Sombras*. Granada: Port-Royal, 2000.
———. "La Estética Cuántica de Xaverio." *Xaverio. Estética Cuántica. Petrales 1997–2000 (Catalogue)*. Granada: Caja de Granada, 2000, 5–10.

———. "Un Nuevo Paradigma Estético." *El Faro*, 29 January 1999, 33.
———. "Principio de Incertidumbre: Consiliencia." *Ideal* 28 July 1999, 17.
———. "La Transgresión del Camino Literario Cuántico." *Debats* 56 (Summer 1996): 72–80.
Morales Lomas, Francisco. "Gregorio Morales y la Estética Cuántica." *Papel Literario (Diario Málaga-Costa del Sol)* 28 July 1998, v.
Murphy, John W. and Jung Min Choi. *Postmodernism, Unravelling Racism, and Democratic Institutions*. Westport, CT: Praeger, 1997.
Peñas-Bermejo, Francisco J. "El Asedio a los Límites en la Poesía de Rafael Guillén." In *Los Estados Transparentes*, ed. Rafael Guillén. Granada: Pretextos/Diputación, 1998.
Plata, Francisco. "El Cadáver de Balzac." *Calas* 5 (June 1999): 225–227.
Rae, Alastair. *Quantum Physics: Illusion or Reality?* New York: Cambridge University Press, 1986.
Salcedo, José Enrique. "Apostillas a la Estética Cuántica." *El Faro*, 11 February 2000, 36–37.
Serrano Caldera, Alejandro. *Los Dilemas de la Democracia*. Managua: Hispamer, 1995.
———. *La Unidad en la Diversidad*. Managua: San Rafael, 1993.
Shlain, Leonard. *Art and Physics: Parallel Visions in Space, Time, and Light*. New York: Morrow, 1991.
Talbot, Michael. *The Holographic Universe*. New York: HarperCollins, 1991.
Villena, Fernando de, ed., *La Poesía que Llega: Jóvenes Poetas Españoles*. Madrid: Huerga y Fierro, 1998.
Zamudio, Arturo. *La Crisis de las Naciones*. La Rioja: Mogila Ediciones, 2000.
———. "En Defensa de lo Diferente." *El Arca*, no. 23–24 (December 1996): 14–15.
Zohar, Danah. *The Quantum Society: Mind, Physics, and a New Social Vision*. New York: Quill, 1994.

Index

Albright, Daniel, on the nature of poetry, 53
Arellanes, Antonio, 23, 26
Aspect's experiment: and non-local action, 17; relevance for quantum aesthetics, 3
Anthropic principle: definition of, 14; and reality, 38
Archetypes, definition of, 21
Aristotle: and dialectic, 174; and logic, 4
Arnold, Mathew, on culture, 169
Art: and Avant pop, 104–105; and beauty, 91; and fuzzy logic, 97; impressionism, 96; meaning of, 71; and philosophy, 72–74; and the physical world, 92–93; and physics, 91–93, 95–96, 113–114, 121; and postmodernism, 103; surrealism, 105; and transcendence, 119; and uncertainty principle, 97; and the world of dynamic fields, 82–85; and the world of eternal return, 77–80; and the world of identity, 74–77; and the world of perspective and sequence, 80–82, 99

Aspect, Alan, Aspect's experiment, 3, 14
Auster, Paul, and anthropic principle, 14
Avant pop, definition of, 104–105
Axelrod, Lawrence, 23, 26

Bachelard, Gaston, on complexity, 134
Barthes, Roland: on the nature of the text, 38; on objectivity, 181
Bateson, Gregory, on knowledge, 157
Baudelaire, Charles, on cosmic unity, 15
Baudrillard, Jean, on the nature of the social, 173
Beck (Hansen, Beck), and avant pop, 104
Bell's theorem, 3
Bergallo, Graciela Elizabeth, 26, 127
Blake, William, 22; and Cartesianism, 94; and art, 101
Block, Alexander, 22
Bohm, David: on implicate order, 7; subatomic particles, 151
Bohr, Niels: on complementarity, 48, 49; and quantum theory, 97, 182

Borges, Jorge Luis: and holographic principle, 17; on religion and philosophy, 12; and synchronicity, 15
Bose-Einstein condensate: definition of, 100; and reality, 102
Breton, André, 22
Broglie, Louis-Victor de, on the nature of light, 48
Bruno, Giordano, 12
Buber, Martin, I and thou relationship, 176
Busse, Thomas, 26

Caro, Manuel J., 27
Caro, María: on the relationship between art and science, 120–121; on uncertainty, 121
Carroll, Lewis, *Alice in Wonderland*, 25
Cassirer, Ernst: *Filosofía de las Formas Simbólicas*, 113; on myth and Symbols, 114
Ceballos, José Gabriel, 26, 127; and quantum aesthetics, 10
Cervantes, Miguel de: Don Quixote, 18; and individuation, 18
Cézanne, Paul, 92; and cubism, 96; and seeing reality, 98; and space, 97
Chomsky, Noam, transformational grammar, 39
Chopra, Deepak, 49
Circlot, Juan Eduardo: and anthropic principle, 12; and fuzzy logic, 22; and holographic principle, 14–15; and imagination, 21–22; and quantum beauty, 25
Collective praxis, definition of, 177
Comte, Auguste: hierarchy of knowledge, 47, 113; on the nature of social reality, 167
Contreras, Miguel Ángel, on poetry, 42
Cuixart, Modest, 22
Cunningham, Merce, on dance, 91

Dalí, Salvador, 105
Davies, Paul: on determinism, 25; on reductionism, 11
Decentered polity, definition of, 172–173
Degas, Edgar, and Cartesianism, 96
Deleuze, Gilles, on nomadism, 173
Descartes, René: and dualism, 93–94; on realism, 42, 93
Deutsch, David, on the nature of reality, 120
Diéguez, Miguel Ángel, 26; on alienation, 53; and anthropic principle, 25; *Los Días del Duopolio*, 52; *En la Gran Manzana*, 25, 43, 53; on the nature of a text, 43; on realism, 52
Double-coding, definition of, 171
Dualism, 3, 5, 7–8, 13, 81–82, 93–94; and alienation, 169–170; and causality, 156; and colonization, 148–149; and culture, 168, 182; and the decentered polity, 171; and destruction of imagination, 159; and essentialism, 175; and the individual, 168; and old metaphysics, 172; and multiculturalism, 185; and political realism, 166–167, 170; and political repression, 166; and reductionism, 155, 157; and social order, 166–167, 183–184; and the social system, 168–169
Duchamp, Marcel, and Avant pop, 104
Durkheim, Emile: on cultural diversity, 184; on facts, 171; reality *sui generis*, 171; on social reality, 167
Dvorac, Mihaela, 26
Dychtwald, Ken, on the nature of the universe, 152

Eddington, Arthur: and dualism, 7; on mindful matter, 39, 172
Einstein, Albert: and dualism, 97; and individuation, 129; on the nature of light, 91, 92, 95; on the nature of time, 95–96; theory of relativity, 7
Eliade, Mircea: on the nature of the sacred, 116; and reductionism, 153; symbolism of the center, 152

Fagerlund, Mikael, 119–121; on cultural liberation, 120
Fagundo, Ana María, on the nature of poetry, 57

Fanon, Frantz, on essentialism, 175
Flaubert, Gustav: on indeterminism, 11; on relationship between art and science, 9
Foucault, Michel, 103
Foundationalism: and alienation, 169–170; and assimilation, 169; and the centered polity, 172; and culture, 168; definition of, 166; and essentialism, 175; and the individual, 168; and old metaphysics, 172; and multiculturalism, 185; and political action, 120; and political realism, 166–167; and social imagery, 168
Franz, M. L. von, 12
Frayn, Michael, *Copenhagen*, 13
Fromm, Erich, on alienation, 167, 169
Fuzzy logic: and art, 95–96, 97; definition of, 6–7; and human identity, 19, 36, 130; and language use, 36–37; and literature, 52–53; and psychology, 130, 132; and quantum aesthetics, 117; and space, 151

García Viñó, Manuel: *El Puente de los Siglos*, 50, 51; and quantum aesthetics, 57; on temporality, 50
Gebser, Jean: on dualism, 173; on the nature of order, 173
Ghyka, Matila C.: *Estética de las Proporciones en la Naturaleza y en las Artes*, 113; relationship between art and science, 113
Goleman, Daniel, on emotional intelligence, 13
Gómez García, Pedro, on reductionism, 156
Gómez-Martínez, José Luis, 49
Gorostegui, Rosario de, 26; on creativity, 41
Green, Brian: elegant universe, 7; unified field theory, 99–100
Grimaldi, Horacio Ejilevich, on Jungian theory, 128
Guillén, Rafael: on anthropic principle, 54; *Los Estados Transparentes*, 55; on holism, 55; *Límites*, 54; on poetry, 54; on ramification thesis, 54

Hawking, Steven, and unified field theory, 92, 99
Hawthorne effect: definition of, 5, 38; and quantum language, 36
Hegel, Georg, on dialectic, 174
Heidegger, Martin: on being, 100; and onto-theological tradition, 166; and thrown-ness, 165; and unconcealment, 101; on Van Gogh's paintings, 101
Heisenberg, Werner, uncertainty principle, 13, 48, 97, 113, 120, 171, 182, 183
Hierro, José: on anthropic principle, 57; on individuation, 57
Hjelmslev, Louis, on language, 35–36
Hobbes, Thomas, 167
Horkheimer, Max, on metaphysics, 181

Ibsen, Henrik, and archetypes, 21
Imbelloni, José, and Toba culture, 150
Impressionism, definition of, 90
Individuation: and creativity, 133–134; definition of, 18–19, 128–129; and language use, 36–37, 133, 140, 141; and literature, 42, 52; and the other, 138–139; process of, 130; and synchronicity, 137–138

Jaffé, Aniela, 12
James, Henry, 22
Jencks, Charles, on double-coding, 171
Jiménez, Julio César, 128
Joseph, Brian, 3
Joyce, James, 22; *Ulysses*, 24
Jung, Carl Gustav: and archetypes, 21–23; and collective unconscious, 41–42; and individuation, 18–19, 128, 175; and quantum physics, 12–13; relationship to Pauli, 12–13; and symbols, 41; and synchronicities, 19–20, 155, 158; and the unconscious, 38

Kant, Immanuel: on autonomy, 43–44; on judgment, 72; on noumenon, 147

Kieslowski, Krzysztof: *Blue, White,* and *Red* trilogy, 20; synchronicity, 20
Kristeva, Julia: on intertextuality, 40; on the nature of a text, 39–40

Lacan, Jacques, 103
Leris, Michael, 147
Lesser, Wendy, 91
Limit syndrome, definition of, 3
Lovecraft, H.P., 22
Lupu, Coman, 26
Lyotard, Jean François, on local determinism, 171

Machado, Antonio: and positivism, 4; on time, 130
Magical realism, 9
Magritte, René, 105
Majork, Sussana, 26
Man, Paul de, 103
Manet, Édouard, on the nature of reality, 96
Mantero, Manuel, on poetry, 56–57
Marcuse, Herbert: on affirmative culture, 168; on one dimensionality, 169
Martínez Manzano, Carlos, 112
Mature democracy, definition of, 177
Medem, Julio: *Los Amantes del Círculo Polar*, 15, 17, 20; and archetypes, 23; and beauty, 24; and synchronicity, 15–16, 20; and temporality, 17
Merleau-Ponty, Maurice, on the social construction of meaning, 186
Métraux, Alfred: Mataco myths, 150; on the nature of fiestas, 149
Mindful matter: definition of, 7–8; example of, 39
Miró, Joan, 22
Monet, Claude, on the nature of light, 91, 92, 95, 96
Monteagudo, Andrés, 26; on the nature of space, 115; on the relationship of art and science, 114–115
Morales, Gregorio, 112; antropic principle, 38, 52, 172, 178; *El Cadáver de Balzac*, 26, 50, 112, 127; *La Cuarta Locura*, 51; on fuzzy logic, 52; *La Individuación*, 42–43; on individuation, 42, 122–123, 139, 174–175; on morphogenic fields, 6, 171; and Newtonian world-view, 82; *El Pecado del Adivino*, 52; and quantum aesthetics, 106; on quantum poets, 58; on symbols, 41
Morin, Edgar, 154, on rites, 156
Morphogenic fields, description of, 6
Multiculturalism: definition of, 185; and quantum literature, 49
Murphy, John, 27

Natural laws, definition of, 1–2
Nerval, Gérard de, 12, 22
Newton, Isaac: on calculus, 47; on laws of nature, 94; Newtonian world-view, 99
Nicolau, Joan, 26; and individuation, 124
Nietzsche, Friedrich: and individuation, 129; on values, 177
Nomad society: definition of, 173–174; and essentialism, 174; and the individual, 174–175; and social imagery, 175–176, 183–184

Ocampo, Estela, on cultural expressions, 149
Orwell, George, *1984*, 53

Parsons, Talcott: Hobbesian problem, 176; on the nature of social order, 167
Pauli, Wolfgang: and anthropic principle, 8; and complementarity, 18; and quantum aesthetics, 12
Penrose, Roger, on reductionism, 11
Pérez Mercader, Juan, on art, science, and intuition, 113
Picasso, Pablo, 12
Planck, Max, on Planck's constant, 48
Plata, Francisco, 26; on alienation, 58; and nonseparability, 16, 24; on poetry, 42; and synchronicity, 19
Poe, Edgar Allan, 22

Index

Positivism: and absolutism, 145; description of, 3; and dualism, 3, 146; and limit syndrome, 3; and reductionism, 149, 153; and signs, 40

Postmodernism: definition of, 103; and quantum aesthetics, 181

Quantum Aesthetics, 9; and archetypes, 21, 43; and avant pop, 104–105; and beauty, 24–25, 102–103; and critique of ideology, 178; and culture, 147, 176, 179; definition of, 10, 28, 93, 98, 111–112, 125; and the democratization of culture, 179; and determinism, 98, 185–186; and diversity, 176, 184; and dogmatism, 112; and dualism, 11–12, 54, 94, 100, 119, 146, 182; and E-m@ilfesto, 112; and engaged research, 187; and epistemology, 17, 159, 171, 183; and fantasy, 120; and fuzzy logic, 117; and grounded metaphysics, 181; and holism, 103, 159; and holographic principle, 76, 152, 184; and human growth, 92, 105, 117; and human identity, 36, 42, 52, 174; and the humanities/sciences divide, 10, 114–115, 181–182; and the implicate order, 106, 151; and impressionism, 95; and individuation, 122–123; and language, 25, 36, 43; and local determinism, 171; and morphogenic fields, 114, 157, 171; and multiculturalism, 185; and nature of writers, 44; and nonseparability, 80, 115, 153; and philosophy of art, 98; and physics, 113; and poetry, 41–42, 53–58; and postmodernism, 103–104, 171, 181; and praxis, 182; and quantum position, 119; and realism, 22, 40, 44, 50, 171, 166–167, 182; and reality, 16, 21–22, 51, 56; and social imagery, 183–184; social philosophy of, 178; and space, 115–116; and surrealism, 105; and symbolism, 40–41; and synchronicity, 155, 186; theory of ramification, 55; and transcendence, 119; and truth, 101, 113; and uncertainty, 54–55, 97–98, 115, 119–121, 171, 182–183; and the unconscious, 38–39, 41–42, 55; and unified field, 99–100, 105; and the world of dynamic fields, 85; and the world of eternal return, 79–80; and the world of identity, 76–77; and the world of perspective and sequence, 82

Quantum Aesthetics Group: brief history of, 26–27, 111–112; E-m@ilfesto of, 26–50, 112, 120; formation of, 26; and philosophy, 73, 127; and politics, 178; promotion of diversity, 176

Quantum anthropology: and appropriate methodology, 159; and critique of dualism, 159; and culture, 147; definition of, 147; and epistemology, 159; and fuzzy logic, 151; and holographic principle, 152, 154–155, 197; and the implicate order, 151; and liminal entities, 158; and morphogenic fields, 157–158; inseparability, 152–153; and the other, 158, 177; and positivism, 146; and reductionism, 149, 153; study of colonization, 148; and symbolism, 156; and synchronicity, 155–156, 158; Toba worldview, 149–151; and uncertainty, 146

Quantum Language: and archetypes, 43; and the human element, 36–37; and individuation, 133, 140–141; and Julia Kristeva, 39; and morphogenic fields, 43; and poetic language, 37–38; and Saussurian theory, 37; and symbolism, 40–42, 114, 156; and Tel Quel, 36, 38; and text, 37–40; and traditional linguistics, 35–36; and the unconscious, 38–39

Quantum literature: and alienation, 53; and complementarity, 54; definition of, 59; and engaged language, 50;

and fuzzy logic, 52–53; and identity, 52; and multiplicity of reality, 56; and nature of the text, 40, 53; and novel, 50–53; and poetry, 53–58, 94; purpose of, 50; and realism, 59; and reality, 51; theory of ramification, 55; and uncertainty, 57, 159; and the unconscious, 55

Quantum physics: acausality, 5–6, 19–20, 158, 186; anthropic principle, 5, 11, 13–14, 36–37, 38, 54, 172; and art worlds, 72, 97; and Aspect's experiment, 3, 17; beauty and truth, 7, 24–25, 102–103, 113; and Bohr, 48; complexity, 6; and culture, 147; and determinism, 98; and dualism, 7–8, 12; and education, 49; and Einstein, 48; and epistemology, 133, 159, 171; fuzzy logic, 6–7, 22, 36, 113, 117; and Heisenberg, 48; holism, 15; and holographic principle, 7, 76, 130; and human healing, 49; and human identity, 13, 17, 18–19, 43; and implicate order, 7, 21, 57, 72, 151; and individuation, 18–19, 36, 122–123, 128–130, 174–175; and legal studies, 28; and management science, 49; morphogenic fields, 6, 114, 171; and multiculturalism, 49; nonseparability, 5, 80, 115; and parallel universes, 16; principle of complementarity, 3, 48, 54; principle of uncertainty, 3–4, 48–49, 95, 98, 100, 113, 119–121, 146, 171; and representational thesis, 44; and symbolism, 114; and traditional physics, 47–49; ubiquity, 6, 14

Quantum politics: and collective praxis, 177; and critique of ideology, 177; and critique of political realism, 167–170; and the decentered polity, 172–173; and democratization of culture, 179; and double-coding, 171; and embodied order, 176–177, 183–184; and facts, 171; and individuation, 174–175; and local determinism, 171; and mature democracy, 177; and nomad society, 173–178; and political realism, 171; and self-legislation, 173; and social realism, 171; and traditional images of order, 165–167

Quantum position, definition of, 119

Quantum Psychology: and creativity, 133–134; and deconstruction, 130; and epistemology, 133; and existentialism, 129; and fuzzy logic, 130; and individuation, 128, 130, 131, 134, 140; and the other, 138–139, 158; and science, 131–132; and socialization, 139–140; and synchronicity, 132

Richard, Dick, 49
Rothko, Mark, on painting, 90
Ruiz de Almodóvar, Agustín, 26; on consciousness, 116; and fuzzy logic, 117

Sartre, Jean-Paul: collective praxis, 177; on engaged literature, 49–50
Saussure, Ferdinand de: on signs, 32; on symbols, 41
Schlick, Moritz, on positivism, 4
Schokkelé, Luc, 26
Schrödinger, Erwin: and anthropic principle, 27; on nature of light, 48; on parallel universes, 5
Serrano, José Luis, on quantum law, 28
Serrano Caldera, Alejandro, on cultural diversity, 185
Shakespeare, William, 22
Sheldrake, Rupert: on morphogenic fields, 6, 21, 154, 157; *Seven Experiments that Could Change the World*, 6
Shlain, Leonard: and archetypes, 23; *Art and Physics*, 92, 96; art worlds, 73; on consciousness, 100; on the relationship between art and physics, 92–93; on the relationship between art and science, 10
Socialist realism, 8–9
Socrates, on unexamined life, 43
Superstring theory, 5

Index

Surrealism, 8, 165
Synchronicity: and causality, 186; description of, 19–20, 137

Talbot, Michael, on the nature of consciousness, 154
Tapies, Antoni, 12
Teilhard de Chardin, Pierre: on art, 94; on consciousness, 91, 100, 103
Thompson, Francis, non-local action, 5
Trakl, George, 22
Truth: and beauty, 7; and uncertainty, 13, 15
Turner, Victor W., on liminal entities, 158

Uceda, Julia, on language, 55–56

Villena, Fernando de, 26; on consciousness, 58; on the nature of poetry, 58; on new poetry, 58

Wheeler, John, 100

Wilshire, Bruce, 49
Wilson, Edward O., on relationship between art and science, 10
Wilson, Jennifer, 26, 27
Wrong, Dennis, on oversocialized conception of man, 167

Xaverio (Javier Bullejos), 112; and anthropic principle, 14; *Colores para Pasear*, 14, 119; on light, 119; non-local action, 17; *Petrales*, 17, 119; and quantum beauty, 24; on transcendence, 119; on uncertainty, 119

Yunduráin, Francsico J., on social implications of quantum theory, 9

Zadeh, Lofti, fuzzy logic, 6–7
Zamudio, Arturo, 26
Zohar, Danah: and Bose-Einstein condensate, 102; *The Quantum Self*, 100; *The Quantum Society*, 98, 100

About the Contributors

Graciela Elizabeth Bergallo is an assistant professor of sociology in social communication at the Universidad Nacional del Nordeste (Argentina). She received a master's degree in anthropology from the Universidad Nacional de Misiones. She has done research on rites, symbolism, the social and cultural impact of research projects, and the "indigenous question," and has published extensively in these fields. She has participated in the compilation of regional poetry entitled *Trópico Sur Antología* (*Southern Tropic*), 1994.

María Caro is an Andalusian artist who has belonged to the Quantum Aesthetics Group since its creation. She has a bachelor's degree in art from the University of Granada, the city in which she currently lives. She has shown her work in Galería Edurne (Madrid), Bubión, and Granada. She has also participated in multiple exhibitions and art fairs throughout the world and has been given various awards.

Manuel J. Caro is an assistant professor of sociology at Barry University. He has published several pieces that deal with democracy and globalization. His work on quantum aesthetics includes his doctoral dissertation, which incorporates an exploration of the consequences that quantum mechanics may have for criminological theory.

Juan Antonio Díaz de Rada holds a Ph.D. in computer science and artificial intelligence. Through his work in cognitive psychology, he has delved into issues related to artificial intelligence, and his dissertation deals with the need to integrate art and science in order to study the processes of knowledge acquisition and to understand the limits inherent in transferring these processes to machines. His current interest relates to the various modes of symbolism.

Mihaela Dvorac holds a bachelor's degree in Spanish filology from the University of Bucharest. She is currently teaching Spanish at the University Spiru Haret (Bucharest). Besides being an instructor of Spanish, she is also pursuing a doctor's degree in comparative literature. She has published several articles and, while participating in the national reform of high school level–education, coauthored the curriculum used in Rumania for teaching Spanish.

Algis Mickunas is Professor of Philosophy at Ohio University. He studied at the Universities of Chicago, Cologne, Freiburg, and Emory. His main interests are contemporary European philosophies, comparative civilizations and philosophies, hermeneutical systems, and semiotics.

Andrés Monteagudo has a bachelor's degree in art (painting and sculpturing) from the University of Granada. He has participated in numerous exhibitions and international art fairs. He has exhibited his work in the Galería Edurne (Madrid) and the Universidad de Granada, and after receiving the National Award for Young Artists he has shown his work in traveling exhibitions throughout Europe.

Gregorio Morales holds a bachelor's degree in Romanic filology. He has directed the Tertulia de Creadores in the Círculo de Bellas Artes in Madrid, one of the institutions associated with postmodernism in Spain. In 1994, he cofounded and presided over the writers group known as the "Salon de Independientes." He has published six novels—including *La Cuarta Locura* (*The Fourth Madness*) (1989)—and several volumes of short stories. In 1998, he wrote *El Cadáver de Balzac* (*Balzac's Corpse*), in which he presents the basic ideas of quantum aesthetics.

John W. Murphy is Professor of Sociology at the University of Miami, Florida. He received his Ph.D. from Ohio State University. His present interests include social philosophy, sociological theory, and cultural studies. At this time, he is involved in the study and promotion of a globalization *from below*.

Francisco Javier Peñas-Bermejo is Associate Professor of Spanish at the University of Dayton, Ohio. He is the recipient of the 1989 Excellence in Teaching Award at the University of Georgia and the 1999 College of Arts and Sciences Outstanding Scholarship Award at the University of Dayton. He is author of several books and numerous articles and has given lectures in the United States, Puerto Rico, Nicaragua, and Spain.

Jennifer Wilson received a master's degree in English literature from San Francisco State University and has taught at the Academy of Art in San Francisco and in Granada (Spain). She has written several articles on the subject of contemporary Spanish poetry and theory, including Rafael Guillén, Gregorio Morales, and quantum aesthetics.